# KNOW YOUR MODEL AERO ENGINES

## SECOND EDITION

by

### R.H. WARRING

A complete guide to
types, performance,
maintenance and operation
— with a chapter on marine
engines

## ARGUS BOOKS LIMITED

Argus Books Ltd
Wolsey House
P.O. Box 35
Hemel Hempstead
Herts HP2 4SS
England.

First published 1979
Second edition 1982

© R. H. Warring 1979
© Argus Books Limited
ISBN 0 85242 819 7

Printed in Great Britain by
Biddles Ltd, Guildford, Surrey

# CONTENTS

# PREFACE TO SECOND EDITION

THERE HAVE been some notable changes in the model aero engine world since this book first appeared in 1979 and this edition has been enlarged and revised to include them. These include the appearance of four-stroke engines as standard productions from several manufacturers—slower and quieter running engines than two-strokes, and particularly suited for powering scale models. Also the demand for larger non-racing type engines for powering very big models has seen the industrial 'chainsaw' type of engine adapted for model aircraft.

Whilst glow engines remain the normal choice in all standard model sizes, interest in diesels has been revived by glow-to-diesel conversions, with surprising results in many cases. Conversions of this nature have proved the suitability of diesel operation for model engines up to .60 cu. in. (10cc) size or even larger, whereas the previous practical limit for diesels was thought to be about 5cc. Also the performance of such diesel conversions can prove quite outstanding—as well as running on a lower cost fuel.

Similarly, there is an interest in converting standard glow engines to spark-ignition, although this only shows advantages in the larger engine sizes. The advantages, again, are operation on a lower cost fuel and better flexibility of speed control. Many of these conversions, too, take advantage of modern technology in incorporating electronic ignition.

Interest in the $CO_2$ engine has, disappointingly, not been as high as expected, or the potential of this power unit for small models warrants. Part of this was due to the fact that some of the first of the new-generation $CO_2$ engines (the type first appeared some forty years ago) suffered from structural problems, mainly because of the use of plastics instead of metals for some vital components. These difficulties have now largely been resolved, and there have also been some quite outstanding developments in improving the efficiency of $CO_2$ engines. Longer, more sustained power runs are now readily possible, using ingenious tank designs.

Besides updating other sections, this new edition also includes *tuned pipes* —originally developed for use on control line speed models to boost engine power output at peak operating rpm. From these 'peaky' pipes have developed other types more suitable for general use, since they are capable of providing silencing as well as power boost. They are coming to be used on contest-type radio controlled aerobatic models, and also on some chainsaw engine conversions which, in their original form, have a relatively lower power output for their displacement, compared with modern model glow engines.

Above: The Irvine 40 is the first of the 'new-generation' glow engines with Schnuerle porting to be developed and manufactured in Britain.

# CHAPTER ONE

# INTRODUCTION

THE OBJECT of this book is to provide easy reading and non-complicated facts that will help any modeller get the best out of his model engine(s) — and dispel the 'mystery' of any technical terms and descriptions he may come across in model journals and other literature. To make it easy to use it has been kept reasonably short, and the whole subject of model engines has been split up into a large number of separate chapters. For immediate information on any particular subject, simply turn to that particular chapter.

Coverage has been confined to the two main types of model engines — glow engines and diesels — together with a brief description of how $CO_2$ engines work (they are simple enough not to require more than this to understand them).

Glow engines predominate, being produced by dozens of different manufacturers throughout the world. They fall generally into three categories —

A typical imported range of Japanese glow engines in air-cooled and water-cooled versions, sizes from .049 cu in to .60 cu in.

sports engines, 'racing' engines and so-called R/C engines.

Sports engines are the 'general purpose' type, suitable for powering free flight models, control line models, and model powerboats (in water-cooled marine versions), where absolute maximum performance is not the aim. In

Above left: The diesel originated in Continental Europe, with Britain becoming a major manufacturer during the 1950s. Early diesels were of long-stroke sideport design (see Chapter 3 for an explanation of these technical terms). Above right: Typical form of British diesel, as it finally evolved. Diesels remain popular in Britain, in sizes up to 2.5 cc capacity (equivalent to .15 cu in).

Above left: The spark-ignition engine was the original type of model engine, rendered obsolete by the appearance of glow engines and diesels in the 1940s. Above right: The glow engine derived directly from the spark-ignition engine, though, and still has a 'family' resemblance.

other words, models operated for week-end sport, not for serious contest work. They are usually easy to learn to start and adjust, and generally non-critical in their behaviour. Size for size, too, the products of major manufacturers can be expected to give a fairly similar performance. But most of them still need running-in, a suitable matching propeller (or waterscrew) size, a suitable fuel and a suitable glowplug. That information you will find in the separate chapters on Running-In, Matching Props, Fuel and Glowplugs.

'Racing' engines are invariably designed to run at higher speeds. The higher the 'peak' rpm the better, as far as power output is concerned (see Appendix 2 for an explanation of this fact!). They are the choice for free flight contest models, control line team racers, and high speed power boat enthusiasts.

Manufacturers continue to vie with each other to get even more power out of an engine of given size for contest work — and here some individual manufacturers are more successful than others. Study the contest results in model journals to see what engines come top in their class. 'Squish' heads and 'Schnuerle' porting have contributed to recent improvements in 'racing' engine performance, as well as other design and construction features such as Dykes rings and ABC construction. All these features are simply explained in

The 'classic' layout of the larger high-performance glow engine, originally evolved in America over a quarter of a century ago. The layout has been hard to improve upon, except in detail design — and notably porting.

Chapters 12 and 14. Also fuels and glowplugs are even more important with these engines.

R/C engines are designed for use on radio controlled models — the main difference being that they are fitted with a carburettor which incorporates a true throttle or speed control (see Chapter 13). Some also incorporate 'racing' engine features for maximum performance. Others are derated in the sense that they are designed to 'peak' at a more moderate speed (to minimise wear and vibration, and also to allow larger diameter props to be used). Fuels, glowplugs (R/C engines need a special type), tanks, silencers and pressurised fuel systems are closely related subjects for getting the best out of an R/C engine.

Diesels really fall into the category of sports engines — slower revving than glow engines and swinging larger propellers. Only a relatively few models are produced as 'racing' diesels, and even then their performance is not usually comparable, size for size, with a racing glow engine. But there are certain contest fields where they can prove competitive, such as control line combat. Their application as R/C engines is limited, too. Partly because they do not respond so well to throttle control (although this is mainly due to less development in this respect), but mainly because they generate more vibration than a glow motor of comparable power. Also they are not made in the larger sizes demanded by most R/C models.

Quite recently the diesel picture has changed. It was found that the modern glow engine with its deflector-less piston and Schnuerle porting could be converted to diesel running by replacing the glow head with a special diesel head, incorporating a contra-piston for compression adjustments. Results

Only one version of a model Wankel rotary engine is in production, the original design developed by Graupner in Germany, with manufacture by OS in Japan.

in many cases have been quite outstanding—a smooth running diesel with its appreciably lower fuel consumption; good throttle response and idling with an R/C type carburettor; more torque than its glow counterpart, with an ability to swing a larger diameter, higher pitch prop; less exhaust noise; and a lower cost fuel.

Diesel conversions have been applied to glow engines from .049 size up to .90. The benefits of diesel conversion usually show up best in the larger sizes—say .29 cu. in. upwards. They contradict the previously held conviction that the model diesel was not a practical proposition above about 5 cc capacity.

As yet, however, (1982) diesel versions of modern ringed piston glow engines have not appeared as a production item. The world interest in, and thus demand for, diesels is not big enough to justify large-scale production. Specialist firms have appeared, though, offering diesel conversions for stock glow engines, pioneer in this respect being Bob Davis in America who introduced diesel conversion heads on the market in 1979.

**Four stroke engines**

Another recent change has been the increasing appearance of the *four-stroke* model engine as a production item. In the early days of model engine development when spark ignition was the norm, a number of manufacturers did produce four-stroke engines, but they were never competitive in performance with two-strokes, and inevitably heavier. Later individual modellers found that four-stroke engines could be converted to glow ignition, with substantial improvement in rpm (and thus power).

Interest in the four-stroke glow engine was revived in the mid-1970's, mainly because of their quieter running and good potential power output at lower rpm—more realistic for scale model aircraft. A number of leading manufacturers now produce engines of this type.

Apart from the working 'cycle' (two-stroke engines have a 'power' stroke every revolution; four-strokes a 'power' stroke every other revolution—which is the main reason they run slower), the main difference between the two types is that a four-stroke has inlet and outlet valves in the cylinder head. These are operated in correct timing sequence by push rods connecting to a cam on the crankshaft; in some cases from an overhead camshaft driven by a toothed belt from a gear on the crankshaft. The carburettor and exhaust also connect directly to the head and inflow of fuel/air mixture and outflow of exhaust gases are controlled by the opening and closing of the inlet and outlet valves, respectively. They work, in fact, on the same principle as a car engine —although employing glow ignition.

Model four-strokes are not quite like 'full size' four-strokes, however. They still run on a fuel/lubricating oil mixture, like a model two-stroke, and

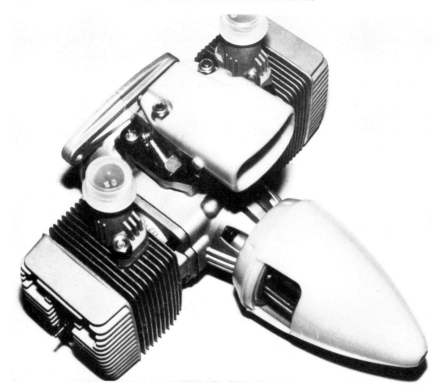

**Large capacity horizontal twin engine—the latest trend in model aero engine development.**

rely on a surplus of this oil finding its way into the crankcase to lubricate the crankshaft bearings, con rod bearings, etc. By comparison, the 'full size' four stroke runs on 'straight' fuel, with lubricant contained separately in the crankcase.

### Multi-cylinder engines

In small engine sizes there is little justification for a multi-cylinder arrangement on a *performance* basis. It will inevitably have more internal friction (the 'rubbing area' of two or more pistons of the same total size as a single piston is greater), weigh more, and be more costly to produce. Theoretically, there is an advantage that with multi-cylinder arrangement reciprocating forces can be more closely balanced, reducing vibration and making for smoother running, but this is difficult to achieve in practice—especially in small engine sizes.

The attractions of the multi-cylinder model engine are, in fact, almost entirely aesthetic. They can be made more compact and so more easily

Scale Bentley rotary aero engine constructed by L. W. Chenery—a superb example of model engineering skill.

accommodated inside a scale-type cowling. In the case of a flat-four con-figuration, or more especially a radial engine, they can also closely approach scale realism in appearance. Basically, therefore, their main application is for scale model aircraft where sheer engine performance is not the main criterion. Many are really outstanding examples of precision model engineer-ing—and often collected as such. But all have one major disadvantage—they are considerably more expensive than orthodox single-cylinder two-stroke engines.

### Prop drives

Another modern trend is the increasing interest in building larger flying model aircraft, demanding bigger and more powerful engines. Standard production sizes for glow engines have, in fact, been increased to .9 cu. in. to meet this need—and even bigger engines are now being used.

The alternative to a big engine is to use two smaller engines, coupled together by gearing to drive a single propeller. This is popularly known as a *prop drive*, the basic form of which is shown in Fig 1.1. The crankshaft of each engine carries a gear, meshing with a larger gear on the prop shaft, which is now quite separate from the engines and is driven by the combined power of the two engines. This will be less than twice the power of one engine due to gearing losses, but with good gear construction this loss can be as low as 10 per cent or less. On a home-built prop drive it could be three or four times higher!

Note with this type of drive that both engines run in the same direction, but the propeller shaft is driven with opposite rotation. Conventional engine rotation is anticlockwise viewed from the front, which means that with a simple prop drive propeller rotation is clockwise, requiring an opposite-hand propeller to normal propeller production.

Fig 1.1.

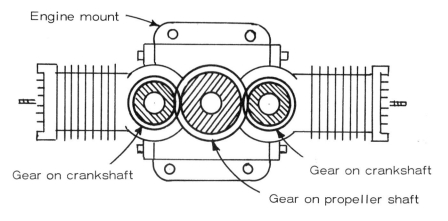

Engine mount

Gear on crankshaft     Gear on crankshaft

Gear on propeller shaft

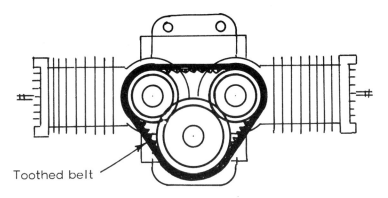

Toothed belt

Fig 1.2.

The object of using smaller gears on the engine shafts with a larger gear on the propeller shaft is to reduce propeller rpm. At the same time torque is increased, so a larger, more efficient propeller can be swung. The trouble is that to reduce propeller rpm to one half or less that of the engine rpm the propeller shaft gear ends up with a large diameter. With the simple gear geometry of Fig. 1.1 this means the two engines have to be mounted with a considerable gap between them.

A much more compact form of prop drive results if toothed pulleys are used instead of gears, connected by a toothed belt (timing belt), as shown in Fig. 1.2. Modern timing belts are quite capable of accommodating the power involved and can operate with an efficiency comparable to or even better than that of good gearing.

### The Big Ones

A prop drive can provide an immediate answer for powering a large model aircraft using two standard, and readily available engines—e.g. two .49 engines giving roughly the power of a .90 cu. in. engine, or two .60 engines giving roughly the power of a 1.2 cu. in. engine. To estimate the actual engine size required, in fact, a very simple rule can be applied. Allow 1 cu. in. engine size for every 16 pounds of model weight. For example, if the model weight is 20 pounds, estimate engine size required as $20/16 \times 1 = 1.25$ cu. in. Two .60 engines coupled with a prop drive should be adequate. (*Note:* in the U.K. maximum permissible total weight for a flying model aircraft is 11 pounds. Bigger models, weighing more, need special permission to fly).

Some of the models now built are quarter-scale, or even larger. Weights may run to 25-30 pounds, even more. Even twin .60 engines with prop drive would find it hard to cope. There is also another point to consider. Prop drives may reduce propeller rpm, but the engines are still screaming away at

Proprietary prop-drive consisting of special mount for two identical engines with integral coupling gear.

Geared mount for two Webra Speed 61 engines, produced as a complete unit by Jack Williams.

high rpm, which is a most unrealistic sound for a scale model—even when silenced.

For that reason, builders of giant scale models have turned to using other types of engines, readily available in larger sizes—a major source being the chainsaw engine. The first of these to find wide acceptance was the Quadra 2 cu. in. chainsaw engine—a simple, rugged two-stroke spark-ignition engine with ample power for a model weighing up to about 24 pounds. (With this type of engine, not highly developed for power performance like a model glow engine, a 'matching' engine size is about 1 cu. in. per 12 lb model weight).

Numerous other engines of this type are available in sizes up to about 6 cu. in., developing as much as 9-10 horsepower. (An average figure for horsepower with a chainsaw-type engine is about 1 horsepower per cubic inch size). They are all spark-ignition engines with a built-in magneto, and mostly fitted with recoil starters. Prices are very realistic for their size—more favourable than model engines, in fact.

All, however, have their limitations—mainly excessive weight and high vibration levels when running. For model aircraft use they can be considerably improved by trimming away excess weight, and rebalancing. Still further improvements are possible by reworking porting, improving carburetion, refining the ignition and raising the compression—and finally fitting a tuned exhaust pipe.

These are jobs for the specialist, with model engineering skills. There are, in fact, small firms now undertaking such work—buying 'stock' chainsaw engines, stripping them right down and then rebuilding them in modified form as engines specifically matched to model aircraft requirements. Unfortunately the amount of individual work required on each engine can double its initial cost.

Industrial - type four - stroke engine fitted with radial mount for model aircraft use. This engine has an overhead camshaft driven by an external timing belt.

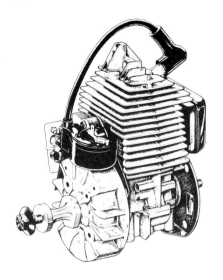

Quadra '35'-a 2.1 cu. in. chainsaw engine adapted to model aircraft use.

### Glow-to-Spark Ignition Conversions

History has turned full cycle in that whereas the modern glow engine derived directly from the original model spark-ignition engine, some modellers are now converting standard glow engines back to spark ignition. Like glow-to-diesel conversions, these, too, have dispelled a number of original beliefs, notably that a glow engine, size for size, is more powerful than a spark-ignition engine. In fact, as a fuel, petrol has a much higher calorific value (and hence more 'power') than methanol, and a spark-ignition engine running at the same *rpm* as a glow engine should develop *more* power. In practice it now seems that the glow-to-spark converted engine is capable of developing about the *same* power, at somewhat lower rpm, the main problem with such conversions being a tendency to overheat. Spark-ignition engines, at equivalent power output size for size, need more *cooling* than glow engines.

With this limitation, the main attractions of glow-to-spark conversion are much reduced fuel consumption, lower cost fuel, and much simpler speed control via spark advance/retard rather than a complicated carburettor. Disadvantages are the greater weight (i.e. the weight of the ignition system) and the possibility of radio interference on a radio controlled model from the points, plug and plug lead. The latter need not prove a particular problem in practice. A 10 kilohm $\frac{1}{2}$ watt resistor in the plug lead close to the plug can provide adequate suppression in most cases, if the contact breaker points are well shielded (e.g. completely enclosed within a metal housing).

The demand for glow-to-spark conversions is relatively small. Most are carried out by individual modellers, but there are one or two specialist firms offering such a service to convert standard glow motors.

Spark conversion of Webra engine with electronic ignition.

Internal details of the Graupner-OS Wankel engine:
(a) rotor.
(b) rotor housing.
(c) internal toothed gear.
(d) fixed pinion (moulded on end of cover).
(e) eccentric shaft.
(f) seals.
(g) inlet port.
(h) exhaust port.
(i) glow plug.

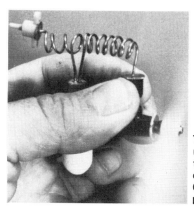

The $CO_2$ engine was another early type, recently revived and put into production in Britain and America. Its chief attraction is that it can be made to work in a very small size (under 0.1 cc, or .005 cu in capacity), suitable for powering very small models.

# CHAPTER TWO

# PARTS OF A MODEL ENGINE

THE EXTERNAL PARTS of a model engine are readily identified — see Figs 2.1 and 2.2. The basic difference in appearance between a glow engine and a diesel is that the former has a plug fitted to the head, whereas the head of a diesel carries a screw fitted with a tommy bar for compression adjustment. Both have the same type of *fuel mixture* adjustment — a needle valve — but a diesel has that extra compression control.

**Fig 2.1.** External parts of a rear-rotary glow engine:

(1) glowplug.
(2) cylinder head.
(3) cylinder jacket.
(4) exhaust stub.
(5) needle valve.
(6) intake tube.
(7) nipple for fuel tube.
(8) crankcase back cover.
(9) crankcase.
(10) main bearing housing.
(11) prop washer.
(12) prop driver.
(13) propeller nut.

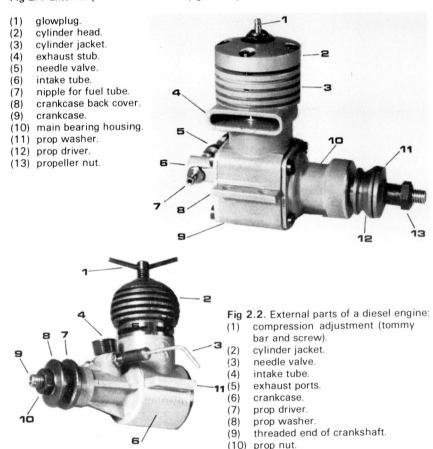

**Fig 2.2.** External parts of a diesel engine:
(1) compression adjustment (tommy bar and screw).
(2) cylinder jacket.
(3) needle valve.
(4) intake tube.
(5) exhaust ports.
(6) crankcase.
(7) prop driver.
(8) prop washer.
(9) threaded end of crankshaft.
(10) prop nut.

Conventional practice is for the crankcase to incorporate lugs for bolting the engine down to bearers or a solid plate when mounting in a model. These are also referred to as *beam mounts*. Some smaller engines do not have beam-mounting lugs but are intended for radial mounting, either via lugs on the back of the crankcase or on the back of a tank bolted directly to the back of the engine.

The exploded diagrams of Figs 2.3 and 2.4 give a more complete picture of the nomenclature of model engines, although not all engines may have the same number of parts. For example, most diesels do not have a locking device for the compression screw (11), nor an O-ring seal on the contra-piston (12). Also the use (or otherwise) of gaskets can differ with different makes of engines. Actual construction can also vary widely (see Chapter 14).

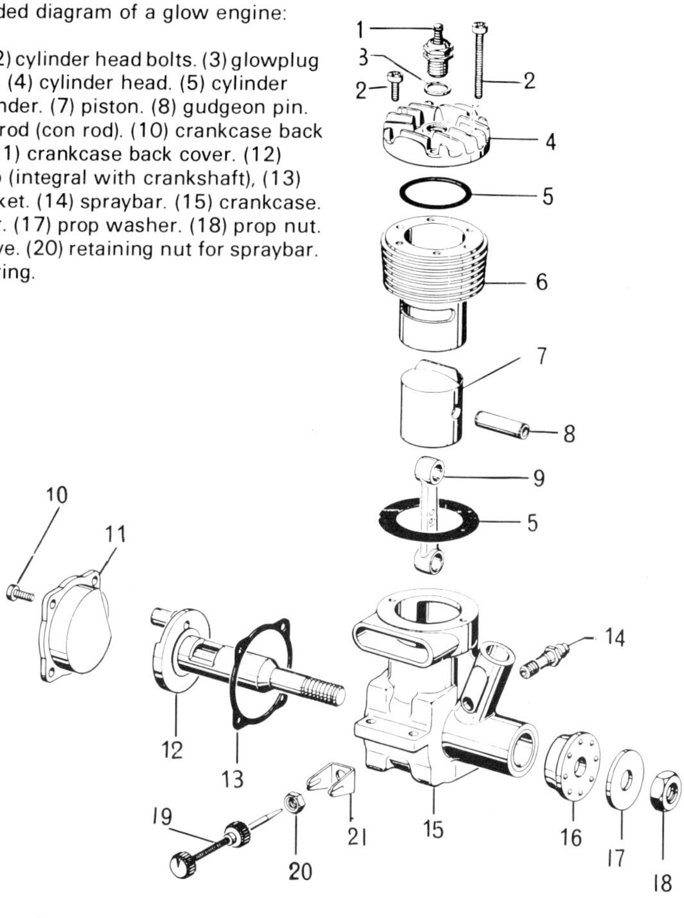

**Fig 2.3.** Exploded diagram of a glow engine:

(1) glow plug. (2) cylinder head bolts. (3) glowplug gasket washer. (4) cylinder head. (5) cylinder gasket. (6) cylinder. (7) piston. (8) gudgeon pin. (9) connecting rod (con rod). (10) crankcase back cover screw. (11) crankcase back cover. (12) crankshaft web (integral with crankshaft), (13) back cover gasket. (14) spraybar. (15) crankcase. (16) prop driver. (17) prop washer. (18) prop nut. (19) needle valve. (20) retaining nut for spraybar. (21) ratchet spring.

## Modern glow engine with R/C carburettor (Super Tigre X-45)

(1)    head
(2)    piston ring
(3)    gudgeon pin
(4)    conrod
(5)    backplate
(6)    crankcase
(7)    rear ball bearing
(8)    front ball bearing
(9)    rear drive washer
(10)   lock cone
(11)   front washer
(12)   prop nut
(13)   O-ring
(14)   needle
(15)   cylinder assembly
(16)   retainer
(17)   gudgeon pin retainer
(18)   pressure tapping
(19)   screw set
(20)   gasket set
(21)   exhaust stack
(22)   R/C carburettor
(23)   glow plug
(24)   crankshaft

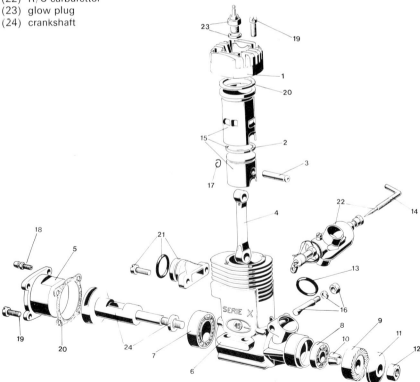

# CHAPTER THREE

# HOW MODEL ENGINES WORK

MODEL DIESELS and glow motors are both 'two stroke' engines. This means that for every single revolution of the crankshaft fuel/air mixture is sucked in (or *inducted*), compressed, fired and exhausted during the time the piston is performing two 'strokes' — one up and one down inside the cylinder. This basic two-stroke action is shown step-by-step in Fig 3.1, but since one cycle of operation overlaps the next one, this diagram depicts more than one revolution to show the engine being started from rest.

At (A) the engine is being turned over (eg, by hand, for starting) and the piston is travelling up the cylinder. This causes a lowering of pressure in the crankcase sucking in a mixture of fuel and air from the carburettor. (In other words, the bottom half of the engine is working as a pump which is sucking.) For clarity, the carburettor is shown on the side of the crankcase with the intake opening to the crankcase in the form of a simple flap valve. The practical arrangement of the intake and the way in which its opening is 'timed' will be described later.

At (B) the piston has reached its uppermost position in the cylinder or Top Dead Centre (normally written TDC). Sucking-in is complete, and the crankcase and cylinder volume below the piston is filled with fuel air mixture.

Continued rotation of the crankshaft, under the inertia of the crankshaft flywheel when starting carries the piston past TDC and on to its downward travel, starting to compress the mixture previously inducted into the crankcase. This pressure closes the intake valve (C).

Compression continues until the piston has almost reached its bottom position (Bottom Dead Centre or BDC). At this point (D) the top of the piston uncovers an opening or *port* in the side of the cylinder connected by a passage to the crankcase. This allows the compressed mixture in the crankcase to escape through this passage and port to the top of the cylinder (above the piston). This port is known as the *transfer port* and the related passage way the *transfer passage*. The compressed mixture cannot escape back through the carburettor since this is now shut off by its valve remaining closed.

At (E) the piston is travelling upwards again (still under the inertia of the flywheel — which is why it is necessary to turn an engine over fairly rapidly for starting). As the piston reaches the transfer port, this opening is closed so that during the remainder of the upward travel of the piston the mixture is trapped in the top of the cylinder and *compressed*.

At some point just before TDC (F) the mixture is ignited, either by the heat of a glow plug element; or in the case of a diesel, by the heat generated by compressing the fuel mixture into the top of the cylinder.

The resulting rapid expansion of gases once the mixture has ignited pro-

**Fig 3.1**

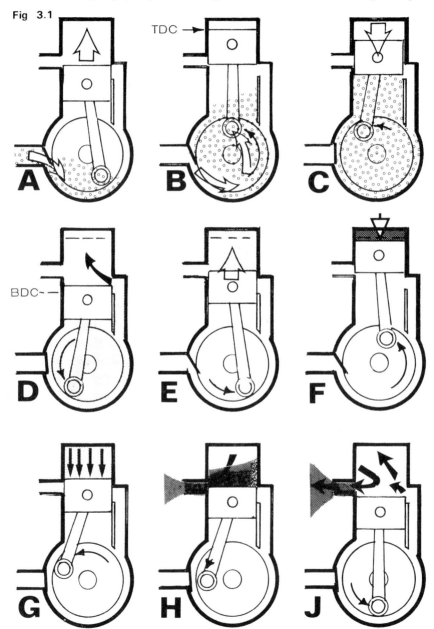

duces a strong pressure on the top of the piston, forcing it downwards (G). (And at this stage the engine should take over and start running on its own, provided it is correctly adjusted.)

Once the piston has travelled part of the way down the cylinder, the *exhaust port* in the side of the cylinder is opened, allowing the burnt gases to escape (H).

Meantime, whilst the piston has been travelling up (previous step F), a new charge of fuel/air has been sucked into the crankcase. Very shortly after the exhaust port has been opened, the *transfer* port is also uncovered by the piston as it continues its downward movement, allowing a fresh charge of fuel mixture to flow into the head (J). Here, in fact, both exhaust and transfer ports are open, the inflow of fuel mixture assisting in blowing out remaining exhaust gases. This is called *scavenging* action.

The cycle of working then repeats itself over and over again — one complete two-stroke cycle being covered by steps E to J. Timing is all-important. The intake valve must open long enough for a 'full' charge to be sucked into the crankcase, and close during the period the mixture is being compressed in the crankcase. The transfer port must open at the right time for this compressed charge to flow into the top of the cylinder through the transfer passage, and close again on the upward stroke to trap mixture in the head.

The mixture must ignite at just the right time. Not too soon as this would oppose the piston trying to reach its TDC position. Not too late or the pressure of the expanding gases will not have their maximum effect in driving the piston downwards, so there will be a loss of power. The mixture itself will take a little time to develop its maximum pressure after it has been ignited, so by firing a little before TDC, maximum pressure is developed *just a shade* after TDC to ensure maximum pressure on the piston *and* continued rotation.

On the downward stroke the exhaust must not open too early otherwise 'driving' pressure will be lost too soon. At the same time it must open early enough and long enough (and must be large enough in area) for all the burnt gases to escape. The transfer must also open *early* enough to promote proper scavenging, and also long enough for the full fuel charge in the crankcase to be transferred to the top of the cylinder. (See Chapter 12 for more on the subject of 'scavenging'.)


*'Two-stroking' and 'Four-stroking'*
The two-stroke working cycle also depends on the *fuel mixture* being correct, so that it is present in the right *proportion* of fuel to air to ignite properly. This will result in 'two-stroking', with the mixture firing once per revolution and the engine continuing to run steadily.

If the mixture is too rich, however (ie, there is too much fuel in the fuel/air mixture), it will not 'fire' on every revolution but on every *other* revolution. This is known as 'four-stroking', recognisable by the engine running much slower and more roughly, with a much oilier exhaust.

Whether the engine 'two-strokes' or 'four-strokes' is governed by the needle valve setting, which controls the actual fuel/air mixture (see also Chapter 13). Of course, if the mixture is *too* rich, it will not fire at all. Equally, if it is too lean (ie, not enough fuel in the fuel/air mixture), the engine will be starved of fuel and stop.

Note that it is the fuel/air mixture *proportions* which determine the 'firing' characteristics of the fuel — not the *quantity* of mixture admitted to the cylinder. The latter is fixed by the 'timing' of the engine and the size of the ports.

**Fig 3.2**

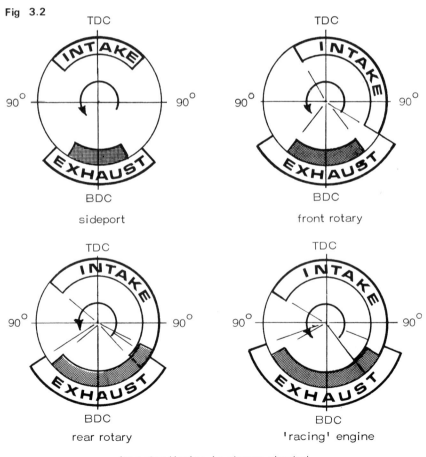

sideport

front rotary

rear rotary

'racing' engine

transfer timing is shown shaded

## Timing

Timing is controlled by port opening and closing. It can differ widely on different types of engines. It is usually described in terms of crankshaft rotation before and after TDC in the case of the intake; and before and after BDC in the case of the exhaust and transfer. Fig 3.2 shows timing for four different types of engines.

**Fig 3.3**

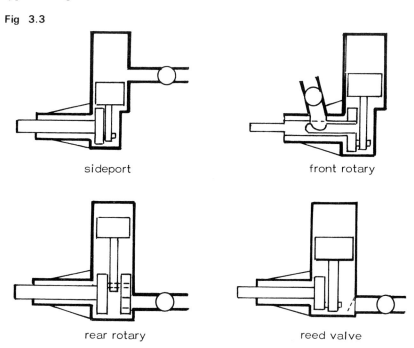

sideport

front rotary

rear rotary

reed valve

## Intake Timing

There are basically four different ways of 'timing' the intake and these are shown in simple diagrammatic form in Fig 3.3. The simplest is to connect the carburettor tube to a port in the cylinder wall which is opened and closed by movement of the piston. This is known as a *sideport* or *three-port* engine. It was widely used on earlier diesels, but is now regarded as obsolete. Its main disadvantage is that timing openings possible are somewhat limited by practical considerations. Timing is also symmetrical, which means that a sideport engine will run equally well with either direction of rotation. For maximum performance, however, the intake port needs to remain open for a longer period after TDC than before TDC, or be *asymmetrical* (when the engine will only run properly in its design direction of rotation).

Most modern engines employ *rotary valves* for intake *timing*. With a *front rotary* engine the carburettor tube connects to a hole in the crankshaft casing.

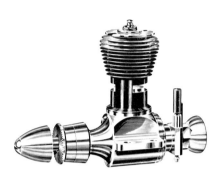

Left: Front rotary induction engine, looking into the open end of the intake tube.
Below: Rear-rotary reed valve engine.

The crankshaft is hollow up to this point, and bored with a radial hole or port which alternately opens and closes the passageway from carburettor to the crankcase. Timing is simply a matter of the circumferential positioning of the crankshaft port and its length.

With a rear *rotary* engine the carburettor tube connects to a port in the rear of the crankcase. Facing the inside of the crankcase back cover is a disc, driven by the crank pin. A hole cut in this disc provides alternate opening and closing of the intake port as the crankshaft (and with it the disc) rotates. Again timing is simply a matter of the positioning and length of the opening in the disc.

In left hand photo, piston is ascending blanking off transfer ports and exhaust. The (rear) rotary valve is opening to allow fuel mixture to be sucked into the crankcase. In second photo, piston is reaching TDC and intake port is closing. On downward stroke, exhaust and transfer ports are opened. Intake port closed.

There are variations on this system — eg, the use of a drum shape rather than a disc for the valve, but the same operating principle applies. There is also a much simpler arrangement where timing is automatically controlled by crankcase pressure (on the same principle illustrated in Fig 3.1). This is the *reed valve* engine where the intake port in the rear of the crankcase is simply covered by a flap of spring metal. This flap opens under crankcase suction, and is closed immediately crankcase pressure becomes positive. Reed valve induction has proved very successful on small glow engines.

### The Four-stroke Engine

Working principle of a four-stroke model engine is shown in Fig. 3.4. The first downgoing stroke of the piston coincides with the inlet valve being opened, sucking fuel/air mixture into the top of the cylinder. At the end of this stroke the inlet valve closes, the piston then moves upwards compressing the mixture into a smaller and smaller volume as the piston approaches top dead centre position. At the appropriate moment the mixture is fired by the glow plug. Note that both valves are closed, so the full pressure of the expanding gases is available to drive the piston downwards. At the bottom of this stroke, the piston reverses direction again and moves upwards at the same time as the exhaust valve is open. Upward movement of the piston thus drives the burnt gases out. At around top dead centre position the exhaust valve closes and the inlet valve opens, ready to draw a new charge of fuel/air mixture in to repeat the working cycle of *four* strokes (down, up, down, up), completed in *two* revolutions of the crankshaft.

Both the two valves and the inlet and exhaust openings are in the head of the cylinder. The valves are normally of poppet type, spring loaded to stay closed and moved down at appropriate times by rockers or cams operated by push rods. The push rods themselves derive their motion from a cam on the

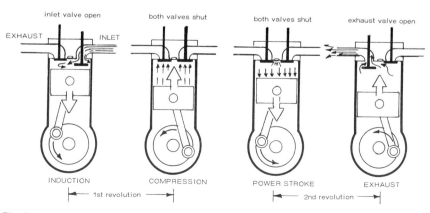

Fig 3.4.

Above left: .35 four-stroke with typical conventional valve-in-head layout with valves operated by rockers and pushrods. Above right: British made 10 cc R.U.E. four-stroke engine with gear driven rotary disc valve in head.

Horizontal twin four-stroke engine for large-scale aircraft.

engine crankshaft. Alternative solutions are to operate the valves directly from a cam on an overhead camshaft, driven at half crankshaft speed either by a toothed belt or gearing; or the use of rotary valves rather than poppet valves, again directly driven. Both these alternatives are comparatively rare.

There are also alternatives to locating both exhaust and inlet in the cylinder head—e.g. at least one model four-stroke engine inducts the fuel/air mixture into the crankcase—but these are even more the exception rather than the general rule.

### The CO₂ engine

The $CO_2$ engine is a special type of two-stroke, working off a supply of compressed gas instead of a fuel mixture. This gas supply is usually carbon dioxide, the chemical symbol for which is $CO_2$, such as provided by Sparklet bulbs.

The operating principle is similar to other two-stroke engines except that the gas is admitted directly to the top of the cylinder. Hence there is no crankcase compression involved, nor the need for transfer ports. The usual method of admitting the compressed gas, and ensuring correct timing, is via a simple ball valve fitted in the top of the cylinder, operated by a spigot on the top of the piston. As the piston approaches its TDC position this spigot lifts the ball to admit a charge of compressed gas. As the piston moves downwards the ball re-seats shutting off the supply until the piston approaches the top of its next stroke — see Fig 3.5.

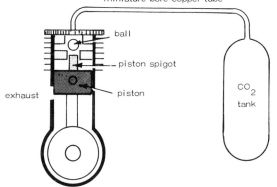

miniature bore copper tube

ball

piston spigot

exhaust

piston

$CO_2$ tank

**Fig 3.5. Basic working parts of a CO₂ engine.**

The timing of such an engine is obviously symmetrical — ie, it will run equally well in either direction. It also depends on the length of time the ball is lifted off its seat by the piston spigot. It is a relatively simple matter to provide adjustable timing, eg, by fitting the crankshaft in an eccentric bush, turning which will alter the actual TDC position of the piston. This, in fact, is an effective method of speed control with a $CO_2$ engine.

Speed control can be a most useful asset. The $CO_2$ engine has the peculiar characteristic that the faster it is operated the more it tends to lose power. This is because the higher the rate of gas consumption the greater the cooling effect of the expanding gas. This can lead to 'icing up' or freezing of the gas, causing partial or even complete blockage. This effect will also be more noticeable in cold weather.

The $CO_2$ motor is particularly suitable for producing in a very small size

Fig 3.6.

**Recharging the $CO_2$ motor tank from a Sparklet bulb. In this model the charging nozzle is fitted to the top of the engine cowling.**

(the fuel is not economic, nor is its performance competitive in larger sizes). Typically the engine is connected to a gas tank which can be recharged several times over from a Sparklet bulb via a separate line terminating in a charging nozzle with a one-way valve — eg, see Figs 3.6 and 3.7.

Characteristic performance of a $CO_2$ engine is a continuously declining power output throughout the run. This is because expansion of the gas cools down the remaining $CO_2$ in the tank, causing the pressure to drop. Starting at normal ambient temperature, in fact, the $CO_2$ tank can cool down to as low as $-4°F$ ($-20°C$) during the run. As a result the gas pressure, which

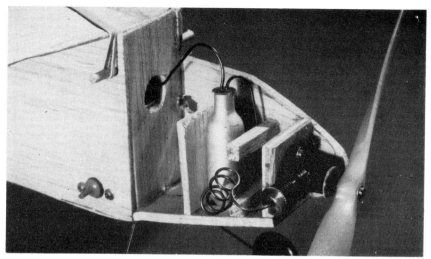

Fig 3.7.

$CO_2$ motor installation showing motor, gas tank and charging nozzle on side of fuselage. The engine is shown mounted sideways (above), but could also be mounted upright or inverted (below left). For best performance the tank should always be mounted vertical, as shown.

Telco $CO_2$ motor has a displacement of only 0.06 cc. Weight complete with gas tank, tube and charging nozzle is ½ oz. It drives 5½ in diameter propeller at 3,500 rpm.

**Telco CO$_2$ engine with conventional tank. Note charging nozzle which can be sited conveniently to suit individual models. Coil of copper feed pipe between tank and cylinder head aids installation but may reduce performance.**

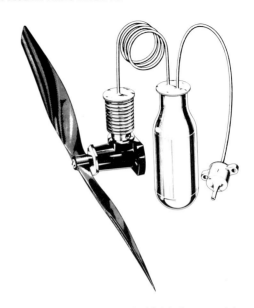

initially may be as high as 600 lb/in$^2$ can fall to about one half this figure, with consequent loss of power.

It also follows that the warmer the CO$_2$ tank to start with the higher the initial pressure, with the pressure loss gradient remaining much the same. In other words a CO$_2$ engine will develop more power from a charge on a hot day than on a cold day. Power, in fact, can be boosted on a cold day by warming the tank in the hand, or in a pocket where it can receive body heat.

There is an ingenious solution to this problem—a double skinned tank with the outer part containing a buffer material which inhibits heat flow *out* of the tank but promotes heat flow *into* the tank. It works in this way. The buffer material melts at about 45°F, which is rather lower than normal ambient temperature. When the tank is charged with CO$_2$ the temperature in the tank falls, to (typically) 40°F, causing the buffer material to freeze. Letting the tank stand for a while allows it to warm up, melting the buffer material which, in melting, will generate latent heat which is stored in the material. If the engine is now run, the tank will start to cool down until the temperature at which the buffer material freezes is reached. In freezing, this material will then give up its latent heat which is conducted *into* the tank to offset the cooling effect. Motor running characteristics as a consequence are modified—a normal initial burst of power gradually falling off, followed by a sustained run of cruising power after the point where the buffer material freezes. (See also Figs 3.8 and 3.9.)

Other developments with CO$_2$ engines have aimed at improving the gas flow into the engine. The very small diameter copper feed pipe is normally

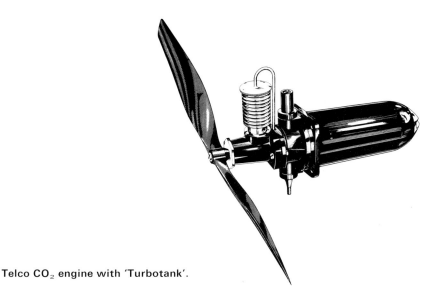

**Telco CO$_2$ engine with 'Turbotank'.**

well overlength, as supplied with the engine. It produces a substantial pressure drop between tank and engine. A considerable improvement in performance can be given by reducing this feed pipe to the minimum length required and keeping the run as straight as possible, with any necessary bends being of generous radius (also making sure that the pipe is not kinked at bends).

Tank design itself can also be improved by incorporating a 'vortex amplifier' or similar fluidic device to provide a supercharging effect, boosting gas flow out of the tank.

A further requirement to make CO$_2$ engines competitive is a customed source of CO$_2$ rather than the standard Sparklet bulb or 'Sodastream' or similar CO$_2$ cylinder. A sparklet bulb only contains 8 grams of CO$_2$ and is an expensive way of buying this fuel. The 'Sodastream' cylinder is heavy and bulky and awkward to handle alongside a relatively delicate CO$_2$ powered model, and the cost of fuel is still substantially higher than that of glow fuels.

**Telco 'Turbotank' 6000 with heat and charger head on CO$_2$ engine.**

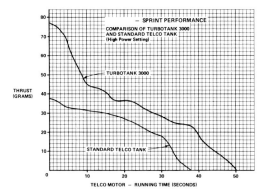

Fig. 3.8

Comparison of performance of Telco $CO_2$ motor with standard tank and 'Turbo-tank' on high power setting.

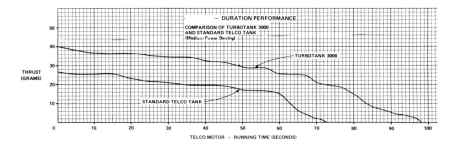

Fig 3.9

Comparison of performance of Telco $CO_2$ motor with standard tank and 'Turbo-tank' on medium power setting.

# CHAPTER FOUR

# STARTING — AND ENGINE CONTROLS

GLOW ENGINES ARE (or should be!) the easiest type of engine to start and adjust since they only have one control — the needle valve which adjusts the fuel mixture. Before attempting to start a new engine, however,

(i)   Read Chapter 5 about running-in a new engine.

(ii)   Mount the engine properly on bearers in a model or on a bench rig (see Chapter 5 on Running-in). Connect fuel tank feed pipe to engine carburettor with flexible fuel tubing.

(iii)   Fit a matching propeller (see Chapter 6).

(iv)   Read the manufacturer's instructions carefully. Make sure that you have a suitable fuel and a suitable glow plug (you should also have a spare glow plug or two).

(v)   You need a battery for starting, matched to the rated voltage of the glow plug (see Chapter 8).

(vi)   You will also need a glow plug clip and leads for connecting the battery to the plug for starting.

The manufacturer's instructions should specify the needle valve setting for

Commercial engine mounts are available in metal or nylon. These can be used for installing an engine in a model, or for mounting on a substantial block of wood to make a test stand for running-in.

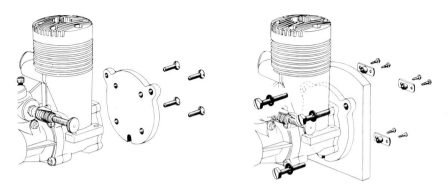

**Method of radially mounting an engine using a special adaptor bolting on to crankcase back cover.**

starting — eg, so-many turns open. Screw down the needle valve until it meets resistance and stops, but do not force tight. Then unscrew the needle valve the specified number of turns.

In the case of engines fitted with throttles (R/C engines), starting should be done with the throttle partly open with a bias to 'slow' rather than 'fast' running. The best position will soon be found by experiment.

Fill the tank with fuel. Cover the open end of the carburettor intake tube with a finger (or thumb) and flick the propeller round a number of times until enough fuel has been sucked up to fill the length of tube between tank and engine. Then give two or three more turns to suck some fuel into the engine for 'priming'. Note that an alternative method of priming is to inject a few drops of fuel directly into the carburettor or through the exhaust port, using an oil can.

Connect the battery to the glow plug and flick the propeller over smartly. It may be necessary to do this several times before the engine fires and starts to run, probably fairly roughly and 'four-stroking'. Close the needle valve slightly until the engine is running more smoothly, then remove the glow clip. Final adjustment can then be made to the needle valve to get the engine running smoothly and strongly. Do not screw the needle valve in too far as this will result in an over-lean mixture and the engine will stop suddenly.

If the engine bursts into life and then suddenly stops, then the mixture is too lean. Try again after three or four 'priming' turns with a finger over the carb intake tube. If it does the same thing, open up the needle valve a quarter to a half turn before trying again.

If there is no sign of the engine firing at all, then there are three likely causes:

(i)   The engine is 'flooded', ie has been primed with too much fuel and/or the needle valve is open too far. Repeated flicking over of the engine will only make matters worse. To clear a flooded engine, remove the plug and blow it dry. Close the needle valve and, with the plug still removed, flick the engine over rapidly a number of times to blow excess fuel out of the engine. If the top of the piston still looks very wet, leave to dry out for a few minutes before replacing the plug.

(ii)   The starter battery is flat, or is not developing enough voltage for the type of plug being used. Always use a freshly charged accumulator (or a fresh dry battery) for starting.

(iii)   The glow plug element is broken. You can check the condition of the starter battery and glow plug by removing the plug, connecting to the battery and seeing if the element glows (see also Chapter 8).

Once you have found the correct 'running' setting for the needle valve, the general rule is that for re-starting from cold the needle valve should be opened up about one full turn (but this can vary with different engines). With practice, and familiarity with the engine, you will soon be able to judge how much 'prime' is necessary for starting (less prime will be necessary if the engine is still hot from the previous run).

Small glow engine installed on a special bench-mount, with glow plug clip attached for starting.

To stop the engine when running, you can either close down the needle valve or squeeze the fuel line to shut off the fuel supply. This will leave the engine 'dry'. Alternatively, placing a finger over the carb intake tube (as in priming) will also cause the engine to stop as it becomes flooded with fuel. This will leave the engine 'wet'. If it is to be restarted again fairly soon afterwards, it should not need priming.

If an engine runs badly, and cannot be adjusted satisfactorily by the needle valve, this could be due to

(i)    The wrong fuel being used.

(ii)   Air leaks (eg, caused by loose cylinder head or crankcase screws).

(iii)  An unsuitable type, or damaged, glow plug.

### Diesels

Starting diesels requires a different technique since two controls have to be adjusted — the needle valve and compression screw. First set the needle valve to the specified starting setting and unscrew the compression screw a turn or two. Suck in a priming charge, just as with a glow engine, and flip the propeller over smartly. It should spin over with little resistance. Screw down the compression screw about a quarter turn and flip over again. Repeat this until the engine shows signs of firing, *but never screw the compression screw down so far that there is a strong resistance to the propeller being turned over.* This means that the contra-piston has been brought too close to the TDC position of the piston, leaving only a small clearance space which could easily fill with fuel and produce a hydraulic lock. To *force* an engine to turn over in this condition could bend the con rod.

Provided the initial needle valve setting is not too far wrong, and is on the rich side, it should not be difficult to find a compression setting on which the engine starts and continues to run fairly roughly. The needle valve can then be screwed in a little. This should produce smoother running, but probably with some misfiring or 'knocking'. Turning the compression screw down a little should cure this, when the needle valve can be screwed in a little more to give smooth running. This may require a further slight adjustment of the compression screw to achieve smooth running.

The basic rule is that the compression should be adjusted to the *minimum* which gives smooth running without 'knocking'. Most diesels will run with *too much* compression but will be labouring under this condition (with higher than necessary stresses developed on the piston, on rod and crankshaft). This can also happen if the needle valve is opened up to richen the mixture when running.

*Laboured* running is a sign of too much compression, which can also cause the engine to slow up and stop. But remember this condition can also be caused by too rich a mixture setting to start with. In either case, the cure is to back off the compression slightly and then, if necessary, readjust the needle valve.

Using fuels containing an ignition additive, the compression setting may have to be readjusted slightly for best running after the engine has warmed up. This is less likely to be necessary on 'straight' diesel fuels, but all diesels have a marked tendency to slow down a little to a steady speed after they have warmed up to normal running temperature.

If the engine starts but only runs in bursts, this could either be too lean a mixture or lack of compression. Try opening the needle valve a little first. If this produces very rough 'four-stroke' running, return to its original position and try increasing the compression a little.

Once running settings have been established on a diesel, suitable *starting* settings usually are: Compression backed off a quarter to half a turn — needle valve opened half to one turn.

These settings are for starting from cold. For restarting a diesel which is still hot, this can usually be done leaving the settings at their running position and merely giving the engine a light 'prime'.

*Flooding* is the chief cause of non-starting with diesels, either by excessive choking to prime, or continual flicking over with the compression backed off much too far. Another possible cause of non-starting is not flicking the engine over smartly enough.

# CHAPTER FIVE

# RUNNING-IN

THERE ARE FEW model engines which do not require 'running-in', preferably before they are installed in a model. In fact, attempting to operate a new engine 'flat out' from the start can lead to serious damage, and even ruin it completely. This is because when new there are inevitable 'tight' spots on the piston/cylinder fit (and between shaft and bearing on plain bearing engines) which will produce excessive friction, and thus heat. This can produce local distortion due to metal expansion, resulting in more friction and more heating — a vicious circle. The object of 'running-in' is to bed down these high spots to produce a really free running engine which will not overheat. It is very much the same with a new car engine, but here manufactured clearances are more generous — so a model engine is really a more critical subject than a car engine.

R/C engines have a fully variable throttle. Use a slow throttle setting for starting, but for running-in operate the engine on full throttle opening and adjust for a rich mixture.

There are three general rules which apply to running-in model engines:

(i)   The smaller the size of the engine, the *less* running-in time it generally needs.

(ii)   Engines with plain pistons normally need *more* running-in time than engines with ringed pistons.

(iii)   Engines with a plain crankshaft bearing may need more running-in than a ballrace engine.

The small ½-A (.049) engines are something of an exception. They are invariably of plain piston type (with plain bearings) but normally require very little running-in. This is largely because there is less metal in small engines and they dissipate heat more rapidly. Just a few runs on a rich mixture (see later) will usually be sufficient — and this can well be done *with* the engine fitted in the model.

Engines in the size range .09 to .19 (and *all* diesels up to 3.5 cc) again normally have plain pistons. They may need a minimum running-in time of 30 minutes, actual running with bursts of maximum speed; and sometimes more.

Larger engine sizes (.29 up to .61) may have ringed pistons, where again 30 minutes (minimum) running-in time should be adequate. Plain piston engines may need as much as twice this running-in time, but not necessarily so. It depends on the engine construction, the precision to which the piston and cylinder have been manufactured, and the actual engine design which governs how equally (or unequally!) heat is distributed through the piston and cylinder.

Ideally every new engine should be run-in on a bench rig, using proper engine mounts bolted to a substantial board or workbench. *Never try to hold an engine in a vice for bench running.* This will inevitably result in irreparable damage — as well as being dangerous. Unfortunately bench-running an engine at home is a messy — and noisy — business. Alternatives are a portable rig; or installing in the model and using that as the 'bench rig'. Either can be taken to a suitable place where noise is no problem.

### The running-in 'Load'

For running-in an aero engine it should be fitted with the same size of propeller as recommended for use with that particular engine for *free flight* or *R/C* (see Chapter 6). If there are alternatives specified, use the largest *diameter* size given (which should also have the lowest pitch). This will give the best 'fan' effect for cooling. Do *not* be tempted to fit an 'oversize' prop in the mistaken belief that slowing the engine down during running-in will be beneficial. It will not — see later.

For running-in a marine engine, a similar 'matching' size of propeller *can* be used for very short runs (not more than one minute at a time), but is *not*

Useful device incorporated in this glow plug clip by Ripmax-MAP is an indicator lamp. Press button and if lamp does not light, the glow plug element is faulty and starting will be impossible.

recommended because cooling of the cylinder head will be very unequal. The proper way to run-in a marine engine is on a bench rig, with a continuous water supply to the cylinder jacket as shown in Fig 3.1. Note that the supply from a suitable water container located above the engine is taken to the *bottom* pipe on the head jacket and the drain tube from the top pipe. This ensures that the jacket is always filled with water.

**Fuels to use**
Use a 'straight' fuel only for running-in. To use a 'doped' fuel is wasteful (because it costs more), and can even be harmful because it may increase the tendency to overheat. The one exception again is the small .049 glow motor which can usually be run-in on its 'recommended' fuel (and may, in some

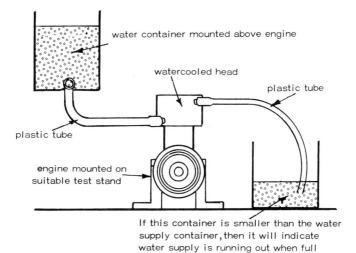

water container mounted above engine

watercooled head

plastic tube

plastic tube

engine mounted on
suitable test stand

If this container is smaller than the water
supply container, then it will indicate
water supply is running out when full

**Fig 5.1**

cases, be difficult to start on a 'straight' fuel). This latter consideration may
also apply to some larger high-performance engines, in which case use a fuel
with the *minimum* of additive necessary to give satisfactory starting and running.

In the case of glow motors, a good 'straight' running-in fuel is 3 parts methanol to 1 part castor oil. Authorities differ on the respective merits of synthetic
oils and castor as lubricants for model engines, so play safe by using castor.
'Straight' fuels with this recommended formulation are available commercially.

For running-in diesels again an 'ordinary' fuel with a high lubricant content should be used. This will not be a 'straight' fuel for it will normally
contain an ignition agent, the proportion of which should be low to medium.
Avoid using 'sports' or 'racing' fuels for running-in diesel engines as the lubricant content will be lower in these formulations.

**Basic running-in technique (Glow engines)**
The first runs should be made with the needle valve of the engine adjusted to
give an over-rich mixture, resulting in 'four-stroking' (see Chapter 3). Several
runs of two to three minutes each should help free any initial tightness, with
the rich mixture ensuring that there is a surplus of lubricant present in the
cylinder.

If the engine is an R/C type with a throttle, the throttle should be kept in
the *fully open* position and ignored. All adjustment is made on the needle
valve.

On subsequent runs, screw in the needle valve until 'four-stroking' changes to 'two-stroking' and the engine speeds up. After a short burst of two-stroking, unscrew the needle valve slightly to revert to 'four-stroking' again. This operation should be repeated — a short burst of 'two stroking', then back to 'four-stroking', increasing the duration of the 'two-stroking' runs gradually, until the engine will run evenly and continuously 'two-stroking'.

In the case of a *plain piston* engine it is best to carry out this running-in procedure in runs not exceeding one or two minutes at a time, allowing two or three minutes between runs for the engine to cool down. With a ringed-piston engine, running could be continuous for convenience, fed from a large tank, with burst of 'two-stroke' running.

There are two important points to watch:

(i)   If the engine shows any signs of slowing up whilst 'two-stroking', then revert immediately to 'four-stroking', or stop the engine and allow it to cool down. Slowing up is a sign of overheating, and the next stage could be seizure and scoring of the bore.

(ii)   Do not adjust for too lean a mixture when 'two-stroking' as this will cause overheating. Just screw in the needle valve enough to make the engine break into 'two-stroking'. Running on an excessively lean mixture can ruin an engine, especially before it is fully run-in.

### Basic running-in technique (Diesels)

Running-in technique for a diesel is basically the same, except that the process will probably be speeded up by using a slightly *larger* propeller than recommended for normal use. Choose a prop of the next *diameter* size up, and next pitch size *down*. Operate basically on a rich mixture, but also back off the compression control as far as practical to maintain 'four-stroking'. To produce 'two-stroking' both needle valve and compression adjustments are used (screw in the compression control until 'misfiring' stops when running on the leaner mixture).

### Judging when running-in is complete

For free flight models, an engine can be considered suitably run-in when it will 'two-stroke' continuously for about half a minute or so with no hint of a change in speed. Longer running-in times may be required on engines for control line or R/C models, when engine runs of several minutes are usually required.

An R/C engine should be checked out by running for the full flight duration anticipated during which run(s) the throttle response can be checked. Any signs of overheating (ie, loss of revs) when running at full throttle can be countered by closing the throttle to richen the mixture and produce a 'cooling down' period. It should pick up to full throttle again without distress after a short slow speed run. If not, then it is probably not yet run-in sufficiently to start operating in a model.

**Above left:** Integral spring starter is fitted to some small glow engines and can be quite effective, especially if used with a nylon prop. **Above right:** Most larger engines — and especially high-speed 'racing' types — are most easily started with an electric starter, powered by a 12-volt accumulator. To use a starter, aero engines have to be fitted with a spinner. Pressing the starter cup fitting against the spinner gives a friction-drive coupling. It is important the spinner be rigid, and firmly secured.

### Can you be too careful?

The answer is, yes! Especially in the case of plain piston engines. These *need* quite a few periods of maximum speed 'two-stroke' running before they can become run-in. Continuous running for long periods at slower speeds may never free up the engine enough for it to run continuously at maximum speed.

### Can you be too careless?

Again, yes! And unfortunately there is always the temptation to regard running-in as something of a nuisance once a model is finished and ready to fly. This can be aggravated by the desire to operate the model on full power from the start, which means using a lean mixture setting. You can get away with very little running-in on some engines, but not on others. And often the engine which has had 'minimal' treatment in this respect (without actually suffering overheating in the process) may have that little extra edge on performance! In other words, two-stroke engines are not for 'coddling'. They produce their best when worked hard. But not after they have been abused.

The most damaging thing for glow engines is an *over-lean* run (i.e. with the needle valve set for an excessively lean mixture). An extended over-lean run (as can occur in careless running-in, or with the engine in use on a control line or radio control model) can ruin an engine.

# CHAPTER SIX

# MATCHING PROPS

THE MAXIMUM speed a model aero engine will run at is determined by the size of propeller with which it is fitted. To achieve maximum *power*, therefore, an engine has to be fitted with a matching size of propeller which permits the engine to achieve the rpm figure at which the engine develops maximum BHP (see also Appendix 2).

In practice, things are a little more complicated. Under 'static' conditions, ie, with the model held stationary, or the engine mounted on a bench rig, a 'matching' propeller will not let the engine develop maximum rpm because the propeller is working as a 'fan' rather than a true propeller. Under flight conditions there will be less load on the propeller, so the engine revs will increase. Thus a correct 'matching' size of propeller has to be estimated on the rpm to be developed *in flight*.

This again will vary with the type of model. Free flight models normally require a fairly generous propeller *diameter* (to develop plenty of thrust), with a moderate to low *pitch*. Control line models generally require a rather higher pitch, with a smaller diameter size to produce the same loading effect

GRP (glass reinforced plastic) or CFRP (carbon fibre reinforced plastic) props are the strongest, but most expensive. They are mainly used on 'racing' engines.

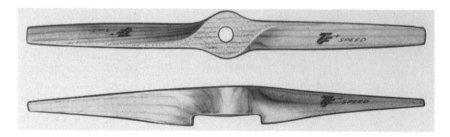

Wood props are easy to rework to trim diameter size and/or reshape blade sections, but are quite vulnerable and easily broken in landings. They may not be safe on high speed engines.

(and thus allow the engine to operate at 'peak' revs). As a very general rule, the higher the peak rpm of an engine the *lower* the propeller pitch it needs on a free flight model: and the *higher* the pitch it needs on a control line model. The amount of 'speeding up' in the air is quite different, too. The higher the pitch of a propeller the more it will 'gain revs' in flight.

Radio control models fall somewhere between the two categories. Generally R/C engines develop their maximum BHP at fairly moderate revs (11-13,000 rpm). This usually means a slightly larger 'matching' propeller size, compared with free flight requirements (virtually a bigger 'load' to hold the engine down to a speed at which it develops peak BHP).

Different 'matching' propeller sizes are therefore usually specified under three separate headings — free flight, radio control (R/C) and control line — for each particular engine. These are sizes found by experience to give the best results when using that particular engine with these three different types of models.

The table given at the end of this chapter gives recommended 'matching' propeller sizes in these three categories for most of the readily available engines. It does not necessarily follow that using a recommended size will give maximum power from that engine. Final performance is also affected by the actual design, type and weight of model — the drag of which is a load on the propeller in flight. But they should be very close to the actual size which will give best possible performance — and certainly close enough to give satisfac-

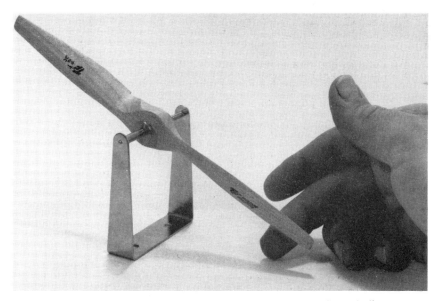

**Proprietary prop balanced (Jim Davis Models) has a self-centering spindle mount to adjust to any size of hole in the propeller hub.**

**A correctly balanced prop will stay horizontal, like this.**

tory results. Some further improvement in performance may be obtained by trying slightly different sizes, remembering:

(i)   A larger *diameter size* will reduce both static rpm and 'in flight' rpm, unless the *pitch size* is also smaller.

(ii)   To get more revs, use a smaller *diameter size* (or reduce the diameter of an existing prop by, say, ½ inch). If this does not appear to produce more thrust from the propeller, try the same reduced diameter size again with a slightly larger *pitch*.

(iii)   Diesels, because they are slower revving generally require larger *pitch sizes* than glow engines of similar size. This is particularly true for control line models. Also because they develop higher torque, they usually require slightly larger *diameter sizes* as well.

Sometimes a change in *make* of propeller can make quite a difference, even if of the same specified size. One reason for this is that the *pitch* dimension specified by manufacturers is often nominal (and may represent a pitch difference of ½ inch or more compared with another make). Differences in blade shape, thickness and section can also affect actual propeller performance.

Nylon propellers are the general choice. On very high performance engines, reinforced plastic propellers are a preferred choice — eg, glass-filled nylon; 'fibreglass' (glass reinforced polyester of epoxy plastic); or carbonfibre reinforced plastic. Wood propellers have the advantage that they are readily reworked, if necessary, to adjust their performance, but are not recommended for use on high speed engines. They are also easily broken, even in an ordinary landing, so they are not more economical than nylon propellers for sports flying.

**Propeller balance**

It is important that a propeller be properly balanced, otherwise it will generate a lot of vibration when running. If a spinner is used, this should also be balanced. The balance of commercial propellers, as manufactured, is not always all that good, and if vibration does show up when running, it is worthwhile checking the prop. To do this it should be mounted on a metal spindle (or tube) which exactly fits the hole in the hub, and the ends of this spindle laid on two horizontal knife edges (eg, two razor blades stuck into a block of balsa — Fig 6.1). File away material from the tip of the heavier blade until the propeller will balance horizontally.

Proprietary prop balancers are also available.

An exactly balanced propeller will not eliminate vibration when running. Although the engine itself is balanced by the crankshaft web being formed to a shape to counterbalance the weight of the piston and con rod, no single cylinder engine can ever be fully balanced.

If marked vibration remains, altering the position of the propeller relative on the crankshaft can sometimes make a difference. Loosen the prop nut, rotate the propeller a quarter of a turn without the shaft being allowed to turn, and retighten. This could reduce vibration — or make it worse! If worse, try another position.

**Fig 6.1**

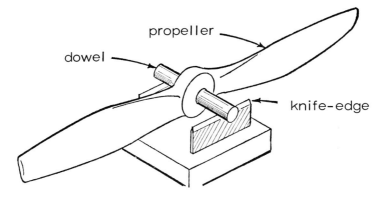

propeller

dowel

knife-edge

## RECOMMENDED PROPELLER SIZES

The following recommendations are based on manufacturers' specifications supplemented by test and operational data. For each propeller size the first figure is the diameter in inches, and the second figure is the pitch in inches. For example, 7 x 4 is a 7-inch diameter 4-inch pitch propeller.

| Engine | Free flight | Control line | R/C |
|---|---|---|---|
| **AUSTRO WEBRA** | | | |
| Webra Speedy | 7×3½   7×4 | 7×5  7×6 | 7×4      7×6 |
| 20 | 9×4  10×3½ | 9×6 | 9×4 |
| 40 | 10×4 | 10×6 | 9×8½  9½×8½  10×6 |
| 61F | — | — | 11×7    12×6 |
| **CIPOLLA** | | | |
| Junior .09 | 7×4 | 7×6 | 7×4  7×6 |
| **COX** | | | |
| Pee Wee | | 4½×2  4¾×2 | — |
| Tee Dee .020 | | 4½×2  4¾×2 | — |
| Babe Bee | | 6×4 | — |
| Golden Bee | | 6×4 | — |
| Tee Dee 051 | | 6×4 | — |
| Medallion 09 | | 7×4  7×5 | — |
| Tee Dee 09 | | 7×4  7×5 | — |
| Medallion 09 R/C | — | — | 7×4  7×5 |
| Medallion 15 R/C | — | — | 7×6  8×4 |
| **DC** | | | |
| Wasp | 5×4  6×3 | 6×4 | 6×4 |
| Rapier | 9×4  10×4 | 8×6  9×6 | 9×4  9×6 |
| Dart | 6×4  7×4 | 6×4 | 6×4 |
| Merlin | 7×4  8×4 | 6×4 | 7×4 |
| Super Merlin | 7×4  8×4 | 6×4 | 7×4 |
| Spitfire | 8×4  9×4 | 7×6  8×4 | 8×4 |
| Sabre | 9×4 | 8×6 | 8×4  8×6 |
| **ENYA** | | | |
| 09 | 7×4    8×3 | 7×6   8×4 | 8×3    8×4 |
| 15 | 8×4 | 8×5   8×6 | 9×4 |
| 19 | 9×4 | 9×5   9×6 | 9×4   10×4 |
| 19BB | 9×4 | 9×5   9×6 | 9×4   10×4 |
| 19X | 9×4 | 9×5   9×6 | 8×6    9×4    9×5 |
| 29 | 10×4  11×4 | 9×6  10×6 | 9×6  10×4  11×4 |
| 29BB | 10×4  11×4 | 9×6  10×6 | 9×6  10×4  11×4 |
| 35 | 10×5  11×4 | 10×6 | 10×6  11×4 |
| 35BB | 10×5  11×4 | 10×6 | 10×6  11×4 |
| 40X | 10×6 | 10×6 | 10×6  11×5  11×6 |
| 45BB | 11×5  12×4 | 11×6 | 11×5  11×6 |
| 60BB | 12×6  13×5 | 11×6  12×5 | 11×7  11×8  12×6 |
| 60X | 11×6  11×7 | 11×6  11×7 | 11×7  11½×7  12×6 |

| Engine | Free flight | Control line | R/C |
|---|---|---|---|

**FOX**

| Engine | Free flight | Control line | | R/C | | |
|---|---|---|---|---|---|---|
| 15 | 8×4 | 8×6 | | 8×4 | 8×6 | |
| 19 | 8×4 | 8×6 | | 8×4 | 9×4 | |
| 25 | 9×4 | 8×6 | 9×6 | 9×4 | 9×6 | |
| 29 | 9×4 | 9×6 | | 9×6 | 10×4 | |
| 36 | 10×4 | 10×6 | | 10×4 | 10×6 | |
| Stunt35 | 10×4 | 10×6 | | — | | |
| 40 | 10×4 | 10×6 | | 10×6 | | |
| 40BB | — | 10×6 | 11×4 | 9×7 | 10×6 | |
| 45BB | — | 10×6 | 11×6 | 10×6 | 11×6 | |
| Eagle 60 | — | 11×6 | 12×6 | 11×7 | 11×8 | 12×6 |
| Hawk 60 | — | 11×6 | 12×6 | 11×7 | 11×8 | 12×6 |
| 78 | — | 11×7 | 12×6 | 11×7 | 11×8 | 12×6 |

**GRAUPNER HB**

| Engine | Free flight | | Control line | | R/C | | |
|---|---|---|---|---|---|---|---|
| HB12 | 7×4 | | 7×6 | | 7×6 | | |
| HB15 | 7×4 | 7×6 | 7×6 | 8×6 | 7×6 | 8×4 | |
| HB20 | 8×5 | 9×4 | 8×6 | 9×6 | 9×4 | 9×6 | |
| HB25 | 8×5 | | 8×6 | | 9×4 | 9×6 | |
| HB40 | 10×4 | | 9×7 | 10×6 | 9½×8 | 10×6 | |
| HB50 | 10×4 | | 10×6 | 11×6 | 10×6 | 11×6 | |
| HB61 | — | | — | | 11×7 | 11×7½ | 12×6 |

**HP**

| Engine | Free flight | Control line | R/C | | |
|---|---|---|---|---|---|
| 40R | 10×4 | 10×6 | 9×8½ | 9½×8½ | 10×6 |
| 40F | 10×4 | 10×6 | 9×8½ | 9½×8½ | 10×6 |
| 61R | — | — | 11×7 | 11×7¾ | 12×6 |
| 61F | — | — | 11×7 | 11×7¾ | 12×6 |

**IRVINE**

| Engine | Free flight | Control line | R/C | | |
|---|---|---|---|---|---|
| .40 Sport | 10×4 | 10×6 | 9×8½ | 9½×8½ | 10×6 |

**K & B**

| Engine | Free flight | Control line | | R/C | | |
|---|---|---|---|---|---|---|
| .35 Series 75 | 9×4 | 9×6 | 10×5 or 6 | 9×6 | 10×6 | |
| Torpedo 40F | 10×4 | 10×6 | | 10×5 | 10×6 | |
| 6.5ccSR11 F1 | 10×4 | 10×6 | | 9×8½ | 9×8½ | 10×6 |
| 61 series 75 | — | — | | 11×7½ | 11×7¾ | 12×6 |

**ME**

| Engine | Free flight | Control line | | R/C | |
|---|---|---|---|---|---|
| Heron | 8×4 | 7×6 | 8×4 | 9×4 | 9×3 |
| Snipe | 9×4 | 8×6 | | 8×4 | 8×6 |

**MERCO**

| Engine | Free flight | | Control line | | R/C | |
|---|---|---|---|---|---|---|
| 29 | 10×4 | | 9×6 | 10×5 | 10×4 | |
| 35 | 10×4 | | 10×6 | | 10×6 | 11×6 |
| 49 | 11×5 | 12×4 | 11×6 | | 11×5 | 11×6 |
| 61 | 13×6 | | 11×8 | | 11×6 | 11×7 |

| Engine | Free flight | Control line | R/C |
|---|---|---|---|

### OPS

| Engine | Free flight | Control line | R/C |
|---|---|---|---|
| 29 | — | 7×7½ | 9×6 |
| 40 | — | 8½×6 | 8½×6   10×5 |
| 60 | — | 8½×9 | 11×7½  12×6 |
| Ursus 60 | — | — | 11×7½  12×6 |

### OS

| Engine | Free flight | Control line | R/C |
|---|---|---|---|
| Pet 09 & Max 10 | 7×4 | 7×5 | 7×6 |
| Max 10F-SR | 7×4 | 7×5 | 7×6 |
| Max 15 | 8×3½  8×4 | 9×4   8×6 | 9×4   8×6 |
| Max 20 | 8×4 | 8×6 | 8×4   8×6 |
| Max 25 | 8×4   9×4 | 9×6 | 9×6 |
| Max 25F-SR | 8×4   9×4 | 9×6 | 9×6 |
| Max 30 | 10×4 | 10×5 | 10×6 |
| Max 35 | 10×4   10×5 | 9×7  10×6 | 10×6 |
| Max 40 | 10×4   10×5 | 9×6  10×6 | 9½×8  10×6 |
| Max 40 SR | 10×4   10×5 | 9×6  10×6 | 9½×8  10×6 |
| Max 60F GP | — | — | 11×6  11×7  12×6 |
| Max 60F GR | — | — | 11×6  11×7  12×6 |
| Max 60F SR | — | — | 11×6  11×7  12×6 |
| Max 60R SR | — | — | 11×6  11×7  12×6 |
| Max 80 | — | — | 11×6  12×6  13×6 |
| FS-60 4-Stroke | — | — | 11×6  12×6 |
| OS Graupner Wankel | 10×4 | 10×6 | 10×6  11×4 |

### TAIPAN

| Engine | Free flight | Control line | R/C |
|---|---|---|---|
| 1.49D | 8×3 | 7×6  8×4 | 8×4 |
| 2.49 Mk4 | 8×6  9×4 | 8×6  7×8 | 9×5  9×6 |
| 19D | 10×4 | 9×5  9×6 | 9×5  9×6 |
| 19D BR | 9×4 | 9×5  9×6 | 9×5  9×6 |

### SUPER TIGRE

| Engine | Free flight | Control line | R/C |
|---|---|---|---|
| G15 | 8×4 | 8×6 | 8×6   9×4 |
| G20/15G | 8×4 | 8×6   7×8 | 8×6   9×4 |
| G20/15D | 8×4 | 8×6   7×8 | 8×6   9×4 |
| X15 RVD | 8×4 | 8×6   7×8 | 8×6   9×4 |
| X15F1 | 8×4 | 8×6   7×8 | 8×6   9×4 |
| X21 | 8×4 | 8×6 | 8×6   9×4 |
| G20/23 | 8×4  9×4 | 8×6   9×6 | 8×6   9×4   9×6 |
| G21/35 | — | 9×7  10×5 or 6 | 10×5  10×6 |
| ST35 | 9×4  9×6 | 9×7  10×5 or 6 | 10×5  10×6 |
| G21/40 | — | 10×6  10×7 | 10×6  11×6 |
| G21/46 | — | 10×6  10×7 | 10×6  11×6 |
| ST51 | — | 10×6  or 7  11×6 | 10×6  11×6 |
| ST56 | — | 10×7  11×4 or 6 | 10×7  11×6  11×7 |
| ST60 | — | 11×6  11×7 | 11×6  11×7  12×6 |
| G60 | 11×6 or 7  12×6 | 11×6 or 7¼  12×6 | |
| G71 | — | 11×7¾  12×6 | 11×7  11×7¾  12×6 |

*Note:* two engines coupled with a prop drive, or very large engines of the chainsaw type, normally swing large diameter, low pitch propellers at moderate rpm. There are no specific propeller sizes that can be recommended as prop size will depend on power available and propeller rpm aimed at. As a general rule, however, with propeller speed of the order of 6000 rpm a propeller with a diameter/pitch ratio of about 0.5 is normally used—i.e. propeller pitch $= \frac{1}{2}$ propeller diameter, approximately. Propeller diameters up to 24 inches may be used with chainsaw-type engines of 6-9 horsepower.

# CHAPTER SEVEN

# FUELS

GLOW MOTOR FUELS are mixtures of methanol and lubricant, with or without additives. Methanol is the basic fuel, and lubricant an essential ingredient. A simple mixture of methanol and lubricant is known as a 'straight' fuel. The usual additive is *nitromethane*, increasing proportions of which produce a progressive increase in rpm and power output, provided the engine is designed to run on nitromethane fuels.

General purpose or sports engines are designed to run on straight fuels. Performance may be improved by nitromethane content up to about 10 per cent, but generally very little thereafter. High performance engines are usually designed to run on specific nitromethane content fuels, and may give progressively increased performance with even higher nitromethane. Generally it is best to use the nitromethane content as specified by the engine manufacturer.

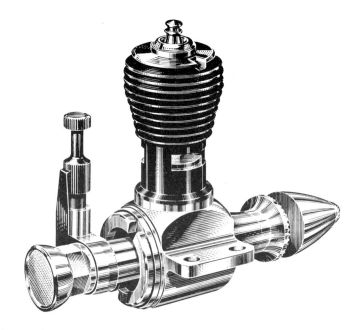

Most glow engines are designed to operate on a fuel with a specified nitromethane content. Those designed for high-nitro fuels, like the Cox, may be difficult to start and adjust on other than the fuel type specified by the manufacturer.

Lacking any specific information on this subject, the following can be taken as a general guide:

(i)   The performance of most R/C engines from .29 to .61 size should be at an optimum with up to 6% nitromethane (maximum). There is little or nothing to be gained by going to the expense of higher nitromethane fuels.

(ii)   With engines in the size range .09 to .25, performance will probably go on improving with nitromethane content up to 15%.

(iii)   Considerably higher nitromethane content is often an advantage with high speed .049 engines (eg, Cox). Consider 40% nitromethane as a maximum, but adequate performance may well be obtained with a fuel having a much lower nitromethane content.

## Straight fuels

A basic straight fuel is 3 parts methanol to 1 part lubricant. This is a good general fuel for running-in new engines, or using in an old engine suffering from wear and loss of compression. High quality castor oil is generally regarded as the best lubricant (or high quality commercial grade castor blends). Synthetic oils are also used by some fuel manufacturers and in 'home mixes'.

Mineral oils do not mix with methanol (except if used with a blending agent, such as a little ether), so are not used as lubricants in glow fuels. Some castor blends contain additives which are insoluble or tend to precipitate out when mixed with methanol. This is the usual cause of white 'fluffy' deposits appearing in a glow fuel in storage.

For rather better performance — and provided the engine has been run-in fully — a straight fuel mixture of 4 parts methanol to 1 part lubricant can be used. This is also the standard specification for an International Competition Fuel, which must contain no additives. (Because nitromethane fuel cannot be used for FAI contests, special competition and racing engines are set up for this type of fuel.)

A number of commercial fuels are available to either of these basic specifications.

## Nitromethane fuels

Commercially produced nitromethane fuels are available with different nitromethane content since no single grade can supply the needs of various types and sizes of engines. In some cases the nitromethane content is increased in simple steps — e.g. 5%, 10%, 15%, etc. According to Model Technics, however, the percentage steps which give a linear increase in power from one blend to another are:

          3%      6%      10%      16%      25%      40%

Fuels with more than 40% nitromethane are not normally used in model engines, except for very special purposes and with the engine specifically set up to match an extra-high nitromethane content.

Nitromethane is a relatively expensive constituent, so lower cost fuels can be produced with alternative additives giving a similar power-boosting effect. As a general rule these fuels are based on reduced nitromethane content, but with further cheaper additive to arrive at an equivalent higher nitromethane rating.

Strictly speaking no other additives should be necessary in a glow fuel, but some are incorporated in commercial fuels to improve other performance characteristics.

Fuels with high proportions of nitromethane additive are generally known as 'hot' fuels (the higher the percentage of nitromethane the 'hotter' the fuel). Conversely, the lower the percentage of nitromethane the 'cooler' the fuel. These designations 'hot' and 'cold' are used in the opposite sense to glow plug heat ranges, ie, a 'hot' fuel matches a 'cool' plug, and vice versa — see Chapter 8.

Nitropropane is an alternative to nitromethane as a 'power' additive but its action is more drastic, producing higher engine stresses and being more prone to pre-ignition. It is seldom used on its own in present-day glow fuels, but may be used as a secondary additive in small proportions.


**Diesel fuels**

A basic diesel fuel consists of a mixture of paraffin, ether and lubricating oil. Paraffin is the actual fuel. Ether itself is a poor fuel, so logically the optimum proportion which should be used is the minimum amount which will give easy starting. Starting and running characteristics can also be improved by an additive amyl nitrite or isopropyl nitrate, further enabling the proportion of ether to be reduced. Only small proportions of such additive are needed as anything more than about 4-5% will have little further effect.

Diesel fuel additive is useful not only to produce a smoother ignition of the fuel but also to reduce the shock loads on the piston when the mixture does fire. They are also referred to as 'anti-knock' additives or *cetane improvers.*

Castor oil, high grade mineral oil (SAE 40) or synthetic oils are all suitable as lubricants, although castor oil is usually preferred. Oil content needs to be one third for adequate lubrication of a new engine when running-in, when a basic diesel fuel of this grade would consist of equal parts of paraffin, ether and castor.

For better performance on an engine when run-in, both the proportion of ether and lubricating oil can be reduced down to about 20% in each case, with additive to compensate for the reduced ether content. For maximum performance the lubricating oil content may be reduced still further, so that the proportion of paraffin in the fuel is proportionately increased.

Most diesels will run satisfactorily on a 1:1:1 mixture fuel with a small proportion of additive (under 2%). For most practical purposes a fuel with 20% lubricant content will give near-maximum performance with all sizes. 'Racing' mixtures, with minimum oil content, are normally intended only to be used with larger sizes of diesels capable of running at speeds in excess of 14,000 rpm. The porting used on many sports type diesels may restrict maximum rpm to lower than this, regardless of the type of fuel used.

The performance of diesels, in fact, is far less responsive to differences in fuel formulation than glow engines, so fuel selection is not particularly important. Any good commercial grade of diesel fuel should suit all types and sizes of diesels.

A point with diesel fuels is that they harden the material commonly used for fuel tubing, to the point where the fuel line can become completely rigid and almost brittle. The only flexible fuel tubing which stays flexible with diesel fuels is neoprene.

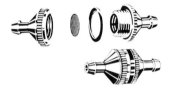

**Typical fuel filter which disassembles for cleaning. This is inserted in the fuel line between tank and spraybar (just cut the fuel tubing at a convenient point and push onto each end of the filter).**

### 'Home Brews'

Most modellers buy ready-mixed commercial fuels. This is generally to be preferred as mixing your own fuels can present hazards due to both the inflammable and toxic nature of many of the constituents. Also effective mixing of constituents can be a problem. (One authority quotes that for hand mixing, shaking a half-filled container, the time required to ensure adequate mixing is 20 minutes!)

However, there are no particular secrets in fuel formulas (suitable proportions have been given in this chapter); and all the necessary constituents are fairly readily obtainable in small or large quantities. The main requirement is to be sure of obtaining high quality constituents, and particularly high quality castor lubricant. Poor quality castor will separate out from a fuel mix in the form of fluffy white lumps after a time. This is particularly likely to occur in colder weather, or if the fuel mix has been standing for a long time. These precipitation products are not harmful, however, provided they are filtered out of the fuel before use.

Methanol used as a basis for glow fuels should be of Hi-Proof quality. This

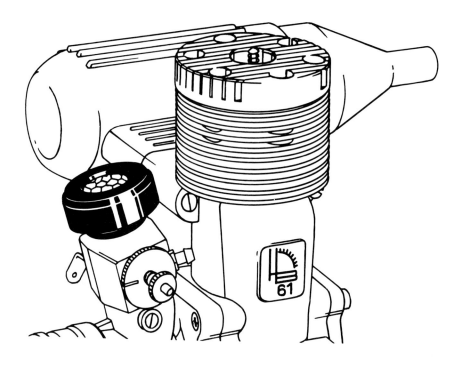

An air filter can be fitted to the intake tube, but this is not really necessary. It could have advantages on an R/C engine with an 'automatic' throttle operated in very dusty conditions, though.

will ensure that it has minimum water content (alcohol absorbs water quite readily, even from the air) and no unwanted additives. For mixing nitromethane fuels, a standard mixture of nitromethane and methanol known as 60% nitromethane can be used. This is actually a 50/50 mixture by volume (60% refers to the nitromethane content by *weight*). The nitromethane proportion can be adjusted downwards, as required, by adding the appropriate amount of methanol. For example, mixing 60% nitromethane with an equal volume of methanol would produce 3:1 methanol: nitromethane mixture.

Ether is sold under various names — Anaesthetic Ether, Ether 0.720, Ether BSS 759, Sulphuric Ether, Ether Meth, Diesel Ether, etc. For practical purposes, these can all be regarded as virtually the same.

# CHAPTER EIGHT

# GLOW PLUGS

A GLOW PLUG is similar in appearance to a spark plug, except that the 'core' of the plug consists of a coil of platinum (or more usually platinum alloy) wire. This is called the element. Platinum has the peculiar property of heating up to 'red heat' when exposed to alcohol vapour (which is what the fuel charge in a glow engine virtually is). This is due to what is called *catalytic* action and takes place with no chemical change in the platinum.

The temperature which a glow plug element reaches depends on a number of factors — the mass of the element, the combination of compression ratio and fuel charge, and the position of the element in the cylinder head. All are factors which have to be taken into account in the design of a glow plug to produce 'firing' of the fuel charge at just the right moment. These are not constant for all engines, so different types of glow plugs are produced to cover various different requirements.

Glow plugs can be categorised broadly in three *heat ranges*, based on the temperatures developed by their elements. A 'hot' plug is one where the element readily develops a high temperature. It is thus mostly used where other conditions present are less favourable to catalytic heating — eg, cold weather

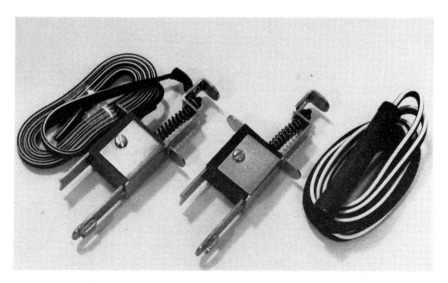

Examples of different types of glow plug clips by Ripmax-MAP.

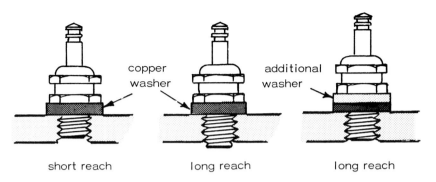

copper washer          additional washer

short reach          long reach          long reach

**Fig 8.1**

operation and/or engines with low compression ratios using 'straight' fuels. A 'cold' plug, on the other hand, develops a lower element temperature to prevent overheating, and thus pre-ignition, using 'hot' fuels in high compression ratio engines. A 'medium' heat range plug signifies a plug with an element designed to meet most average conditions. In other words, a general purpose glow plug, which is often called a *standard* plug.

From the above, it will be appreciated that changing a glow plug for one of different heat rating can be used to adjust 'timing' of a glow engine (particularly when changing from one grade of fuel to another): and also to meet different conditions. A 'hot' plug, for example, would tend to overheat with a 'hot' fuel, causing pre-ignition and also overheating of the element so that it burns out.

The element temperature is also affected by the manner in which it is exposed to the fuel. If well shielded, ie, either by being located well up in the body of the plug, or in the head itself, it will be less likely to be 'wetted' on starting; but by being less exposed to the fuel charge may not heat up so much. In other words it may have both 'hot' plug characteristics (for starting) and 'cold' plug characteristics for running. This is a general characteristic of a plug with a short body length, or *short reach* plug.

A *long reach* plug has a longer body which brings the element down further into the fuel mixture. This will tend to give it 'hotter' characteristics when running, and may also assist starting because of the closer proximity of the element to the bulk of the fuel charge. So the difference produced by a change in plug from short reach to long reach can only be determined by trying the two out. Most engines will take both types of plugs, but *some engines do not have enough clearance between the head and top of the piston at TDC to accommodate a long reach plug.* This should be checked first, otherwise damage to the engine could result. The actual 'reach' of a long reach plug can, of course, be reduced by using an extra plug washer. See Fig 8.1.

The other basic type of glow plug is the one fitted with a projection across the bottom of the plug body which partially shields the element — Fig 8.2. This is known as an *idlebar* plug. Its purpose is to prevent 'raw' fuel being splashed onto the plug element and cooling it off when the engine is running on an over-rich mixture. It is thus specifically applicable to throttled engines as used on radio control models, and so is often called an *R/C plug*. Otherwise it is similar to other types, being made in different heat ranges and with long or short reach.

**Fig 8.2**

**Three different types of glow plus are shown here. The one on the left has an idlebar; the centre plug a heat-sink core; and the one on the right is a standard plug with an unprotected spiral element. Idlebar and standard plugs are the main types.**

There are also numerous variations in individual designs. The most common difference is one of choice of element material with claimed superiority in catalytic action and/or life. There are others with radically different forms of construction. The 'Hotspot' plug    Fig 8.3 — features a soft iron core inside the element, the main purpose of which is to act as a 'heat sink' to help maintain a more uniform element temperature on the basis that this should give improved starting and more even element temperatures under different running conditions. This type of plug does not need an idlebar for use with throttle engines. The more even operating temperature of the plug reduced 'thermal fatigue' and the core also provides mechanical support to resist shocks. Both features contribute to a longer element life.

The GloBee plug differs from other types in having the element wound in the form of a flat spiral rather than a helical coil. It is also supported by high temperature glass seals. This configuration enables the whole element area to be presented to the fuel, with the heat range determined by the position of the element in the body — see Fig 8.4. It is a more expensive form of construc-

**Fig  8.3**                              **Fig  8.4**

**Above left: Close-up detail of the 'Hotspot' glow plug with soft iron core acting as a heat-sink for 'automatic' element temperature control.**
**Above right: Close-up detail of the 'GloBee' plug with spiral-wound element.**

tion, but one which has achieved considerable success with high performance engines using very hot fuels, and in giving long plug life.

### Which plug suits which engine?

Some engine manufacturers specifically produce plugs to match their own engines. Most do not. This is of little significance, as a standard plug will suit most engines, except for high-performance engines operating on high-nitro fuels where a 'cool' plug may be required. There are also the exceptions where a specific engine may only run well on a particular plug element incorporated in the actual engine head itself. In this case it is not possible to fit an alternative — only change the whole head if the element fails. This is only usual in some smaller glow engine sizes.

Any individual engine may, however, run *better* on one type of plug than any other (with further possibilities of experimenting with different heat ranges in that plug type). The main differences between different makes of plugs, fitted to a particular engine, however, will probably be in the *life* of the plug. Some plugs have a much longer life in certain engines, used with the same fuel, than others. This is something that can only be determined by experience. It is worth trying a change in plug type or make if plug life is short as continual replacement of glow plugs can be expensive.

Glow plugs can fail by burning out, or by mechanical failure. The risk of burning out is aggravated by using too 'hot' a plug with a 'hot' fuel; and by leaving the starter battery connected for too long after the motor is running (see notes on Glow Plug Voltage). Mechanical failure can be expected eventually since the glow plug element is subject to a pressure shock on every compression/firing stroke and although it is conventionally in the form of a spring, any metallic spring will fatigue and fail eventually under cyclic loading.

Most glow plug elements are also subject to 'catalytic ageing', although this will vary with different element compositions. This is a progressive loss of catalytic action, making the glow plugs less efficient (effectively making it become slowly but progressively 'colder'). This could be noticed by a gradual loss of engine performance and/or more difficult starting. Some fuels have a noticeable catalytic ageing effect on glow plug elements, notably those containing reclaim methanol or low grade castor lubricant.

### Glow plug voltage

One other aspect of glow plugs has yet to be mentioned. That is the *voltage rating*. This refers to the starter battery voltage as developed across the glow plug (this will generally be a little lower than the nominal voltage of the battery used). The majority of glow plugs are rated as 1.5 volts or 2 volts. There is a risk of overheating a 1.5 volt glow plug using a 2-volt accumulator for a starter battery, unless a dropping resistor is used in one lead to reduce the actual

**Fig 8.5**

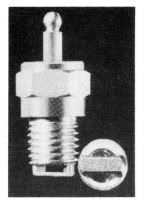  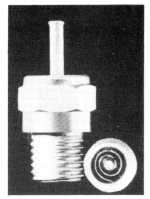

The three versions of the 'GloBee' plug — R/C plug on the left with idlebar.

voltage across the plug to 1.5 volts maximum. Using a 1.5-volt battery (eg, a dry battery) with a 2-volt glow plug could result in the element not heating up enough for satisfactory starting.

Whether the battery voltage is correct or not can easily be judged visually with the glow plug removed and connected directly to the battery. With the correct voltage the plug element should glow bright red. If it is dull red, then the battery voltage is too low, and starting can be difficult or even impossible. If the plug element has a yellow glow, the battery voltage is too high.

Make sure that the plug is dry before trying such a test. A flooded plug just removed from the engine will only develop a dull red glow with the correct voltage, although it will soon dry out to its 'normal' glow.

Increasing battery voltage to overcome stubborn starting is *not* the answer, provided the element has the right colour glow. It is an almost certain way of burning out the element; and will certainly destroy the element if the engine is left running for any time with the battery still connected.

## Glowplug Drivers

A glowplug driver—or glow driver, as it is usually called—is a pocket size solid state electronic device designed to connect to a 12 volt car battery and supply a low voltage (1.5v or 2v) output for connection to the glow plug when starting glow engines. Its basic advantages are that it eliminates the need for carrying and maintaining a separate 2 volt starter battery; also the supply is from a high capacity battery which is unlikely to go flat. Even when supplying high current to a heavy-duty glow plug, current drain on the main battery is unlikely to exceed 1 amp.

Glow drivers are also available with multiple outputs, providing a simple means of starting twin- and multi-cylinder glow engines, i.e. a multi-output glow driver provides independent and isolated outputs to connect to each plug, each output normally having an LED indicator which lights up when connection is made.

Two precautions should be observed when using glow drivers. Never short the output when the driver is switched on (preferably use a glow plug clip which cannot accidentally short if it slips); and always switch off the driver when the lead is removed from the plug. If left switched on, it will continue to draw current from the main battery and the output could accidentally be short circuited (e.g. if the plug clip is dropped into wet grass).

## Starting troubles

Starting troubles are not necessarily due to the wrong grade of glow plug being used. They are mostly caused by a flat starter battery, or a flooded engine. However, a change in glow plug may be indicated to cure troubles when running, eg:

(i)   If the engine loses power when the battery is disconnected and cannot be adjusted satisfactorily by the needle valve, try a 'hotter' glow plug (or a 'hotter' fuel).

(ii)   If the engine pre-detonates ('pinks'), try a 'cooler' glow plug (or a 'cooler' fuel).

Cox engines are unusual in having the glow plug element mounted integral with the head, rather than using a separate glow plug. To change or replace it is necessary to replace the complete head.

# CHAPTER NINE

# FUEL TANKS

ON A FREE flight model the fuel tank can be of any convenient shape and size. Models subject to changes in attitude, such as control line models and R/C models require a tank design which ensures a continuous supply of fuel regardless of whether the model is climbing, diving, looping, flying inverted, etc.

Four different basic designs of fuel tank are shown in Fig 9.1. The first is rectangular in shape with the feed pipe terminating at the rear bottom end. This shape is generally suitable for all free flight models (when it could be square instead of rectangular); and for control line models not subject to violent manoeuvres (eg, team racers).

The second is wedge shaped — the classic form for stunt control line models. The feed pipe is located at the apex of the wedge with the tank positioned so that the apex is on the outside of the circle. Under the centrifugal force

**Fig  9.1**

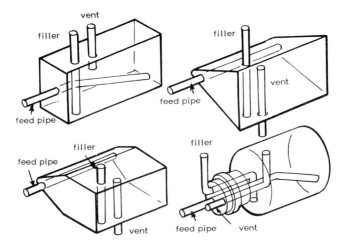

developed in control line flight, fuel is always thrown to that side of the tank, so the fuel pipe will remain submerged until all the fuel is used up. It is also a symmetrical tank, so it will feed equally well in both normal and inverted flight. It is also suitable for both conventional anti-clockwise control line circuit flying, or anti-clockwise circling.

Main disadvantage of the wedge is that it has only half the capacity of a rectangular tank of similar overall dimensions. The more compact form, shown in the third diagram, is thus usually preferred. This is the conventional form of control line 'stunt' tank.

Rectangular and wedge tanks of these three types are made in metal (usually tinplate, but sometimes brass) with soldered seams. Vent and filler pipes are brass, also soldered in place. Internal baffles may also be incorporated in such tanks to reduce fuel surge in climbs and dives. Baffles are usually only considered necessary in large, or long tanks.

For free flight models, a plastic bottle can be used as the basis of a suitable tank — simpler and lighter than a metal tank, and more compact for the same capacity if circular in section. This type of tank is shown in the fourth diagram. It has a metal or plastic cap, in which the three metal pipes are fitted. A metal cap with soldered pipes ensures leakproof joints. A plastic cap needs to be used with a soft rubber washer or takes the form of a rubber bung to provide satisfactory sealing.

The radio control model needs a tank which will continue to supply fuel in every possible attitude. The choice is normally a 'klunk' tank, which is virtually identical to the plastic bottle tank just described but with the rigid fuel pipe terminating just inside the cap. Fig 9.2. The fuel line is then continued with a length of flexible tubing, to the end of which is attached a weight ('klunk' weight). This weight is subject to the same acceleration forces which may cause fuel to be thrown from side to side in the tank, so 'follows' any movement of fuel in a partly empty tank — hence the end of the pipe always stays in the fuel. It is not necessary for the pipe to double forward in a dive for under this flight condition the model will be accelerating, tending to force fuel to the back of the tank.

**Fig 9.2**

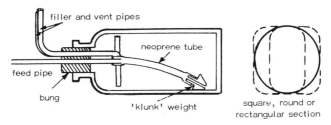

filler and vent pipes

neoprene tube

feed pipe

bung

'klunk' weight

square, round or rectangular section

The klunk tank is very simple in principle and works really well. The main requirement is that the fuel pipe must be flexible, and remain flexible. Many plastics harden and stiffen after being in contact with fuel for some time, so the usual choice of material is neoprene, or silicone tubing. Further refinement in design is possible. For example the klunk weight may also incorporate a filter element, or a separate filter may be fitted to the end of the filter tube. However, there is one disadvantage with a filter fitted inside a tank. It means that the tank must be disassembled to clean the filter at regular intervals. A filter in the fuel line is thus a better proposition since the tank need not be disassembled, once made up, and so the end seal is not disturbed.

The klunk tank will serve equally well for other types of models, including power boats. Special marine tanks are, however, produced, normally made in brass for corrosion resistance.

The fuel tank for a pressurised system need not be a special design, for it is only necessary to supply pressure to either the vent or filler pipe (preferably the vent) and plug the other pipe — see Fig. 9.3.

**Fig 9.3**

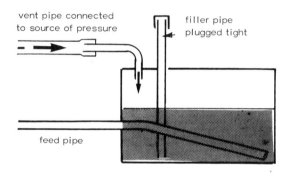

vent pipe connected
to source of pressure

filler pipe
plugged tight

feed pipe

The correct position for fitting a tank is with the top of the tank level with, or very slightly below, the spraybar or fuel jet of the engine carburettor — Fig 9.4. This will ensure that the carburettor only 'sucks' fuel and there is no flow under gravity (as would occur if the tank was above the spraybar) which could cause flooding. Placing the tank top in line with the spraybar will also ensure that there is a minimum change in 'head' of fuel which the carburettor has to suck up as the tank empties.

**Fig 9.4**

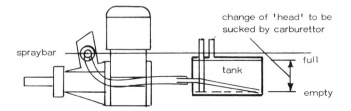

change of 'head' to be
sucked by carburettor

spraybar

tank

full

empty

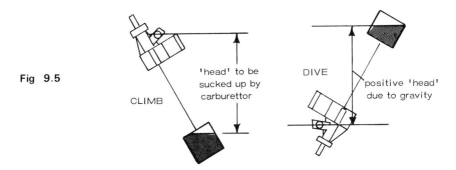

**Fig 9.5**

CLIMB — 'head' to be sucked up by carburettor

DIVE — positive 'head' due to gravity

It is even more important to place the tank as close to the engine as possible to minimise change in fuel 'head' in a climb. Fig. 9.5 illustrates this point. A large change in head must be avoided, if possible, since suction present in the carburettor is fixed by the needle valve setting and engine speed. A large change in head in a climbing attitude could result in insufficient suction being present to overcome that 'head', with the result that the engine is starved of fuel and stops. Conversely, in a prolonged dive, a large positive fuel 'head' could result in excessive fuel being fed to the engine by gravity.

Where this effect is noticed on a model, or where it is strictly necessary to locate the tank some appreciable distance from the engine (as on certain types of scale models), then a pressurised feed may provide the solution.

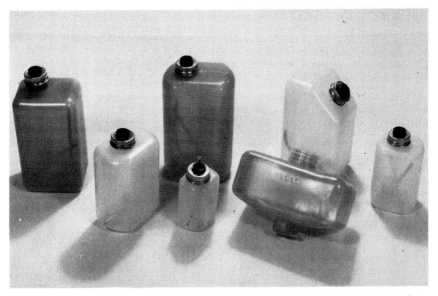

Examples of different shapes of plastic bottle klunk tanks. The tank bottom right is an exception, being a marine tank with bottom feed.

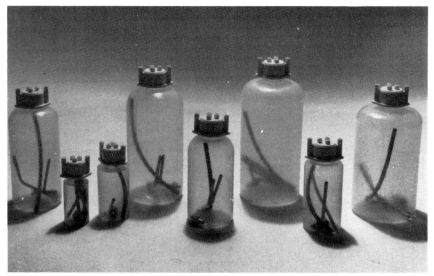

**More examples of klunk tanks based on a cylindrical plastic bottle.**

There are also special tanks designed to provide a constant feed, regardless of apparent 'head' changes. An example is the chicken-hopper tank, shown in Fig. 9.6. This is really two tanks in one, the bottom section having the fuel outlet and filling only to a certain level. At this point it blocks its own vent, so the rest of the fuel used in filling the tank fills the *header* section. As fuel is withdrawn from the bottom tank, its vent is uncovered, allowing fuel to flow from header to the bottom tank until the vent is closed again, shutting off further flow into the bottom tank. In this way the fuel level in the bottom tank is maintained at a constant level, being automatically topped up from the header tank until the header is empty.

This type of tank is still subject to changes in head as in Fig 9.5, so its use is mainly limited to models which do not change their operating *attitude* greatly —eg, control line team race models and power boats.

**Fig  9.6**

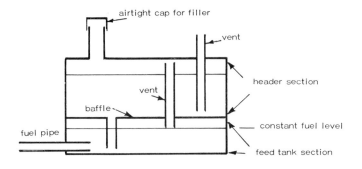

Shaped metal parts and tubes, and soldered-up version on right of a conventional control line stunt tank.

## What size of tank?

Fuel consumption obviously varies with the size of engine, and also with its type, and the rpm at which it is operated. As a general guide, the following fuel consumption figures can be used for glow engines:

*Sports and R/C glow engines operating at 10-13,000 rpm*
Typical fuel consumption figures:
0.1-0.125 ounces/minute per .10 cu. in. displacement.
0.06-0.075 ounces/minute per cc displacement.
*Racing glow engines operating at 15-20,000 rpm*
0.15-0.18 ounces/min per .10 cu. in. displacement.
0.09-0.11 ounces/min per cc displacement.

For example, a .60 cu. in. R/C engine 'propped' to run at 12,000 rpm, could be expected to use $6 \times 0.125 = 0.75$ ounces per minute. Thus an 8 ounce fuel tank would give a duration of $8/0.75 = 10\frac{1}{2}$ minutes as an estimated figure.

In the case of *diesels*, fuel consumption figures can be from $\frac{1}{4}$ to $\frac{1}{3}$ that of glow engines of the same capacity. Actual consumption however can vary considerably with different designs of engines of the same size. The following

figure is quoted as typical of *diesel conversions* of larger size glow engines:
0.1 ounces/minute per .10 cu. in. capacity.
0.06 ounces/min per cc displacement.
Diesel conversions of smaller, high-revving glow engines can show up to 3 times this figure.

*Spark-ignition engines* are normally the most economical of the lot. Here are typical figures for *spark conversions* of glow engines:
2-strokes:   0.033-0.04 ounces/min per cu. in. displacement.
             0.02-0.025 ounces/min per cc displacement.
4-strokes:   0.020-0.025 ounces/min per .10 cu. in. displacement.
             0.012-0.015 ounces/min per cc displacement.
Lower-revving spark-ignition engines can show even more favourable consumption figures.

*Note:* the capacity of fuel tanks may be quoted in fluid ounces or cc. For rapid conversions used the following figures:
1 fluid ounce = 30 cc approx. (actually 28.4 cc).
10 cc = $\frac{1}{3}$ fluid ounce approx. (actually 0.35 ounce).

**Pressurised fuel system installation with tank vent pipe taken to crankcase pressure tapping point on engine (see Chapter 13).**

# CHAPTER TEN

# SILENCERS

THE TWO-STROKE engine is naturally a noisy type of machine, especially when it has an exhaust opening directly from the cylinder. The only way this exhaust noise can be reduced is by taking it through a silencer before it can escape into the air. Unfortunately, to be really effective a silencer needs to be quite large, relative to the engine size, and it does have an unwanted effect on engine performance. Silencers inevitably introduce some resistance in the exhaust flow or 'back pressure', resulting in a loss of rpm. This 'back pressure' also tends to shorten engine life.

Nevertheless a silencer is generally necessary, and *in many cases essential* in order to be able to operate a model engine at all without incurring complaints about noise, or even a total ban on using particular flying sites or ponds.

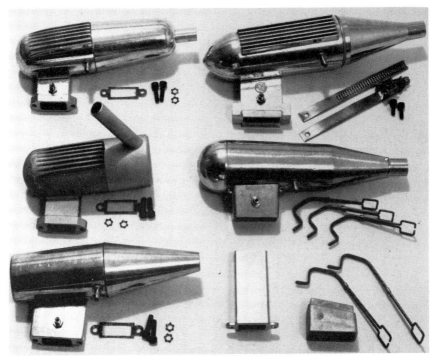

Some examples of different shapes and sizes of engine silencers, together with mounting fittings.

The worst 'offenders' are the high-revving engines. Not only is the *amount* of noise generated directly related to the rate at which a two-stroke fires (ie, its running speed), but also the *quality* of noise becomes more objectionable to people who are not modellers. And here there is really no answer. It is virtually impossible to silence a 'racing' engine down to 'quiet' levels without at the same time getting a drastic reduction in performance. To produce a 'quiet' model engine, in fact, would need a design concept almost exactly the opposite to what most modellers want (and, more important, what most modellers would consider buying). It would be a low-revving engine, of very moderate power output for its size, and with a silencer volume many times that of the cylinder displacement.

'Matching' silencers are now available for every model engine. They are generally satisfactory for sports use. That is, they provide a noticeable amount of silencing without too much loss of performance and not too drastic an effect on engine life. They are the 'answer' as far as the average modeller is concerned, except where he may come up against regulations which demand a greater degree of silencing than that provided by a 'standard' silencer. In this case there are two possible solutions:

(i)   Try fitting a larger silencer — eg, the 'standard' silencer for the next lar-

Conventionally the silencer is clamped or bolted directly onto the stub exhaust on an engine.

gest engine in the same range provided it will fit the engine exhaust properly.

(ii)   Fit another type of proprietary silencer which is designed to provide a better degree of silencing than a typical manufacturer's standard one.

Marine engines are less of a problem as there is usually space to fit two (or more) silencers in series, if necessary, and the extra weight is unimportant.

Where loss of performance cannot be tolerated, then the only answer at the present time is the fitting of a *tuned pipe* instead of a silencer. This is really like a very long silencer, of 'tuned' volume to eliminate back pressure, by 'megaphone' effect, but at the same time producing a substantial noise transmission loss. That means it is a very critical size, matched closely both to the displacement of the engine and its design operating rpm. In fact, both the engine and pipe have to be designed with this matching in mind. It is thus a specialised type of production — and an expensive one to buy.

Apart from high cost — and the fact that a tuned pipe will only match the engine for which it has been developed — a tuned pipe normally tends to make the engine less flexible as regards starting, and particularly running adjustment.

### Basic silencer design

The subject of silencer design could fill a whole book. As far as simple silencers are concerned, however, the basic form is an expansion chamber (Fig 10.1) which works as an elementary 'noise filter'. Length is more important than diameter. The longer it is, the more effective it should be. Its silencing performance can also be improved by lining the inside with some sound-absorbing material and/or extending the pipes into the chamber. The latter, however, will substantially increase power losses.

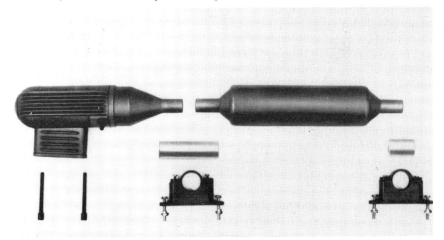

Standard form of silencer with second silencer to be connected in series. Connecting tubes and mounts for second silencer are also shown.

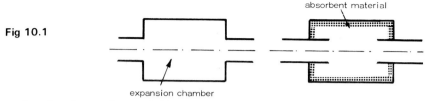

**Fig 10.1**

A rather better form of silencer is shown in Fig 10.2. Here the main pipe passes right through the expansion chamber, but is drilled with holes to allow 'noise' to escape radially and be absorbed in this chamber. In this case, however, the point at which the holes are (ie, the length from the front of the chamber to these holes) is quite critical in determining the silencing effect produced. Other factors which affect performance as a silencer are the number of holes and their actual diameter. It is difficult to design a silencer of this type to produce good silencing at more than one specific frequency.

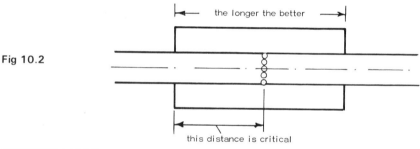

**Fig 10.2**

Silencers fitted to these engines have pressure tap nipples. For a pressurised fuel system, silencer tap is connected to tank vent pipe with plastic tube (and tank filler pipe blocked).

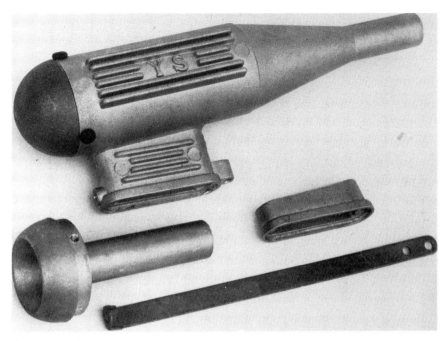

Conventional closed silencer with alternative open-front fitting. Other items in this photo are an extension manifold piece and a mounting strap for the silencer.

Two other variations on basic silencer design are shown in Fig 10.3. Both introduce back pressure, and have certain critical features. With the internal tube type, for example, the longer the tube the more it will reduce noise at specific frequencies, but the more it will tend to let noise escape at intermediate frequencies. And a model engine develops a lot of 'annoying noise' at different frequencies, not just at the frequency corresponding to the speed at which it is running!

Fig 10.3

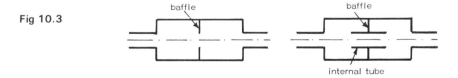

Most model engines have a sideways-facing exhaust, which makes the design of a silencer even more difficult. For simplicity it wants to be bolted directly over the exhaust, which means carrying the exhaust gas flow through a right angle — Fig 10.4.

**Fig 10.4**

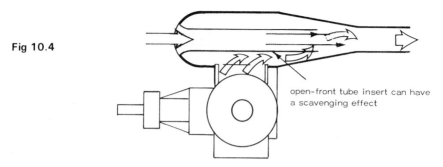

open-front tube insert can have a scavenging effect

The more effective silencers used on full size engines (e.g. automobile engines) normally contain noise-absorbent materials, or are designed on a 'multi-pass' system where the gases are directed backwards and forwards through the silencer. Neither type is particularly suited to model engines which operate at much higher speeds.

Silencer design, therefore, is best left to the expert — or the enthusiast who is prepared to spend a lot of time on trial and error development. Most commercially available silencers are of basic type, with individual ideas added by their manufacturers. Most are of 'closed' type with the front end blanked off (and rounded for streamlining when fitted on an aero engine), with a side-entry exhaust. Inevitably this change in direction of the exhaust gases introduces back pressure, so some silencers of this type are made with an open front, the idea being to create a venturi effect to help reduce back pressure losses inevitably at the expense of some loss of silencing.

Simple silencers are normally made to butt directly over the exhaust stub on the engine and be secured in place with a strap and two screws. This makes fitting a 'matching' silencer simple and easy, but to be fully effective any joints between silencer and engine exhaust stub need to be completely gas tight. Many installations would perform a lot better if assembled with a heat resistant liquid gasket material applied to the mating faces before tightening up in place.

Otherwise, once fitted, a silencer should give no problems, other than requiring regular draining to remove liquid fuel residue which may have collected in the expansion chamber. Some silencers are fitted with a drain plug for this purpose. Some are also fitted with a flap which can be opened to permit a priming charge to be shot directly into the cylinder head for engine starting, but this can be a source of noise escape so is not generally regarded as good practice.

## Tuned Pipes

A tuned pipe looks rather like an oversize silencer. In fact, it is a unit designed to boost the power output of an engine, but at the same time it can provide a

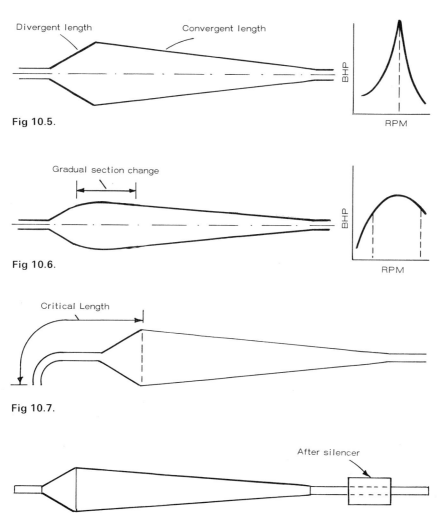

Fig 10.5.

Fig 10.6.

Fig 10.7.

Fig 10.8.

degree of silencing. Basically it comprises a chamber consisting of a short divergent section followed by a longer convergent (or sometimes parallel) section, with the shorter (front) section connected to the engine exhaust outlet. Shapes and volumes need to be matched to a particular engine and if these are right the effect is to cancel exhaust back pressure and reverse it at the right moment, promoting a suction effect in the combustion chamber for more efficient fuel induction. In other words it works in a similar manner to a supercharger which boosts power by 'blowing', but the tuned pipe does the same thing by 'sucking' on the exhaust side.

The design of a tuned pipe is quite critical and normally done on empirical (cut-and-try) lines. Most manufacturers of 'racing' engines now normally produced tuned pipes as accessories specifically tailored to individual engines. They are not something the individual modeller himself attempts to make. Most tuned pipe designs follow similar geometry (differing in volume for different engines), but there are two distinct types.

Pipes with an abrupt change of section where the divergent and convergent cones join are characterised by 'sharp' tuning, i.e. they develop their peak power only at a particular rpm—Fig. 10.5. These are generally known as 'peaky' pipes. If the change in section is more gradual, as in Fig. 10.6, then the tuning is much broader. That is to say, power boosting is produced over a range of rpm rather than at a specific peak rpm. This is a more flexible system, although maximum power produced will tend to be lower than that given by a 'peaky' pipe on the same engine.

A 'peaky' pipe is thus a preferred choice where absolute maximum power is required, but it has its disadvantages. It is more critical on tuning adjustment and relies on the engine running at peak rpm in order to produce maximum power. If engine revs are off, either way, it will deliver much reduced power.

This represents a particular problem where control line speed models are concerned—which is where tuned pipes were first used. Unless the model can accelerate sufficiently to 'offload' the propeller to enable the engine to reach peak rpm, power will remain way down. To achieve this, energetic 'whipping' of the model may be necessary before enough model speed is realised for the engine rpm to reach peak.

Whichever type of tuned pipe is used, final engine rpm realised is, of course, dependent on the propeller size used. For maximum power, prop size would be selected to give an in-flight rpm corresponding to the rpm for maximum BHP for that engine. Fitting a tuned pipe complicates this by modifying the rpm at which peak engine BHP is actually produced so some means of adjusting the tuned pipe performance is necessary.

The critical volume is this respect is that between the top of the piston and the maximum section of the tuned pipe (i.e. where the divergent and convergent sections join. This volume is adjustable by varying the *length* of the front pipe joining the tuned pipe to the engine exhaust outlet—Fig. 10.7.

Standard 'tuning' practice is to first run the engine on open exhaust on the selected prop, checking the rpm. The tuned pipe is then fitted with an over-length front tube and resulting engine rpm measured. The front pipe length is then reduced a little at a time, when each reduction should give an increase in rpm. This is continued until the last length reduction produces a *drop* in rpm. The optimum length for the front pipe is then determined as one or two lengths back from this and a final inlet pipe cut to this size and fitted.

Being a 'tuned' length as far as resonance is concerned, a tuned pipe should also provide a degree of silencing. If necessary, however, additional silencing can be provided by fitting a secondary silencer or after silencer to the tail pipe—Fig. 10.8. Provided this is a 'straight-through' type, this should not affect the 'boost' performance of the tuned pipe.

This system is commonly employed when tuned pipes are fitted to aerobatic radio controlled models giving power boost with extremely silent running. It is important with this type of model that a broad tuned pipe should be used as the engine will be called upon to work over a range of different speeds. A pipe which is too 'peaky' will produce a sudden surge of power at full throttle as the boost cuts in.

Prototype of a small (.21 cu in) four-stroke engine.

# CHAPTER ELEVEN

# ENGINE STARTERS

HAND STARTING, using the forefinger placed against a propeller blade to flip the prop over smartly, is used on all engines from the smallest to the largest size. This is easy enough for right-handers since virtually all model engines run anti-clockwise, viewed from the front, and are started by flipping over right-to-left. More awkward for left-handers, who usually find it easier to flip the 'bottom' rather than the 'top' blade to achieve the right direction of rotation.

It is a knack which is easily learnt, although the inexperienced or clumsy modeller may often suffer a cut finger. If this is a worry, some protection can be given by wearing a finger stall, or a special moulded finger-protector known as a 'chicken finger'. This will only protect the starter finger, though.

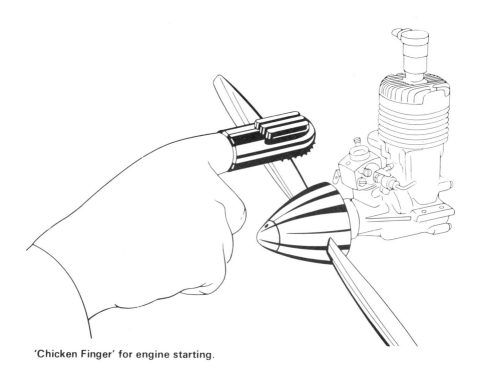

'Chicken Finger' for engine starting.

**Above left: Spring starters are fitted as standard to the Davies Charlton range of diesels from 0.5 cc to 1.5 cc. Above right: Cox (America) fit spring starters to a number of their smaller glow engines.**

There is actually more chance of getting cut fingers by inadvertently reaching through the virtually invisible propeller disc of a running engine to adjust the needle valve.

A self-starter fitted to the engine has obvious attractions, but has definite limitations. It needs to be simple and light, virtually restricting choice to some simple form of recoil spring. Various types of spring starters have been fitted to production engines, but the only type which has survived is the open coil spring mounted behind the prop driver, one end of which is hooked behind a propeller blade or a special backplate, the prop turned 'backwards' one turn or so and released. Disengagement of the spring is automatic once the prop 'overruns' the spring travel. Provided the engine is suitably primed and adjusted, it should start within three or four attempts.

Simple spring 'self-starters' of this type are most suitable for small glow engines. They are less suitable for diesels because of the higher resistance to being turned over (due to higher compression ratios), but are fitted to one production range of diesels up to 1.5 cc size (Davies Charlton). The only production glow motors still produced with spring starters are the smaller Cox models.

A basic feature of a spring starter is that you can use it, or ignore it. In other words, the fact that it is fitted to the engine does not interfere in any way with hand starting.

The more powerful the engine the more painful a mistake in hand starting can be, and the process can become hazardous at times hand-starting a racing engine fitted with a small diameter propeller. The best answer in this case is to use a 'mechanical' starter applied to the propeller from the front of the engine, when the hand can be kept well away from the propeller disc.

**Sullivan Starter**

Early engine starters of this type were purely mechanical design (recoil spring or even hand-cranked), or modified fractional horsepower electric motors (eg, car engine starters). These have given way to purpose-designed electric starters specially developed for use with model engines and produced by such manufacturers as Sullivan (USA), Kavan, Marx Luder and Graupner-Bosch (Germany).

**Marx Luder starter.**

The (model engine) electric starter is designed to take a variety of 'drive' fittings on its spindle. The most common fitting used for starting an aero engine is a rubber friction cup to be pressed over a spinner to achieve 'drive' between starter and engine. An alternative friction rubber may also be available to drive directly against a propeller hub (eg, with the Kavan starter). For starting marine or car engines (fitted with a flywheel), the 'drive' fitting may be a friction wheel to press against the rim of the flywheel or, more usually, a grooved pulley to engage a belt looped around the flywheel starting groove.

For aero engine starting a spinner drive is usually best, and less likely to slip off. A point to bear in mind is that the spinner fitted to the engine crankshaft should be quite sturdy to stand up to the beating it is likely to get from a starter, and also must be securely fixed. The method of fitting some spinners is unsatisfactory as the whole spinner assembly can be spun off by the starter instead of turning the engine over.

Although relatively expensive (and also requiring an accumulator to work them), electric starters offer many advantages for use with all sizes and types of model engines. They are usually designed to work off a 12-volt battery (a separate accumulator or a car battery), but will normally develop enough power to start smaller engines on a 6-volt accumulator.

### In-flight Starters

There are now even in-flight starters — small electric motors mounted on the engine and coupled to a crankshaft gear by a Bendix or similar type drive, working on the same principle as a car engine starter. They are intended for use on radio controlled models, so that if the motor is intentionally (or accidentally) stopped in flight, it can be re-started by a radio signal switching on the starter circuit, which includes its own starter battery. In the case of a glow motor, this will also need a 1.5 volt supply to the glow plug also being

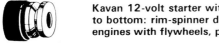

Kavan 12-volt starter with spindle fittings shown separately, top to bottom: rim-spinner driver, prop driver, rim-contact driver for engines with flywheels, pulley drive for belt starting.

switched on at the same time, automatically switching off when the engine starts and the starter disengages.

An in-flight starter can also be used as a starter with the model on the ground, using an external battery and manually operated switch.

# CHAPTER TWELVE

# EXPLAINING ENGINE DESIGN FEATURES

THE SIZE OF an engine is correctly defined by its *displacement* or *swept volume*. This is the volume actually traversed (or 'swept') by the piston, given by the product of the *stroke* (or distance between TDC and BDC) and the area of the *bore* (or internal cylinder diameter) — Fig 12.1. Mathematically, therefore, $\qquad$ Displacement = .7854 x stroke x (bore)$^2$

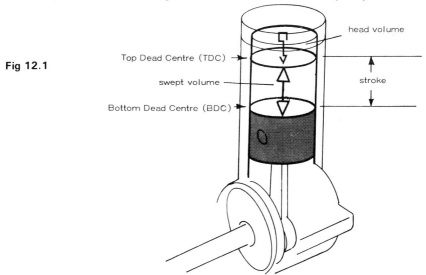

**Fig 12.1**

head volume

Top Dead Centre (TDC)

swept volume

Bottom Dead Centre (BDC)

stroke

To call this volume the *capacity* of an engine is not correct. When the piston is at the top of its stroke there is a small volume remaining in the cylinder head, called the *head volume*. *Capacity* is equal to displacement *plus* head volume. This difference between capacity and displacement is significant only in determining the *compression ratio*, ie,

$$\text{Compression ratio} = \frac{\text{capacity}}{\text{head volume}}$$

$$\text{or} \quad \frac{\text{displacement + head volume}}{\text{head volume}}$$

Glow engines normally have moderate to high compression ratios — eg, from about 6:1 to 10:1 — chosen to match the grade of fuel they are designed to run on. Compression ratio can be changed by fitting an additional or

Example of a modern high-performance front-rotary R/C engine.

thicker head gasket or washer to lower the compression ratio; or by the fitting of an alternative head with a different head volume.

Diesels have a *variable* head volume, altered by the position of the contra-piston. Thus they have a variable compression ratio. The typical running setting for a diesel gives a compression ratio of about 25:1, or even higher. This means that the clearance between the top of the piston and the bottom of the contra-piston is very small at TDC.

### Bore and stroke

An engine with the same dimensions for the bore and stroke is called a 'square' engine. If the stroke is substantially greater than the bore, it is a *long stroke* engine. If the stroke is less than the bore, then it is a short stroke or *over-square* engine — see Fig 12.2.

**Fig 12.2**

Note: to scale drawn these three engines have the same displacement

A

B

C

long stroke or
under-square

square

short stroke or
over-square

The significance of this is that the shorter the stroke the less the distance travelled by the piston per stroke and thus the lower the *piston speed*. This is an advantage for high speed running. Also it enables the overall height of the engine to be reduced, and the crankcase diameter, resulting in a more compact layout.

The disadvantage of a short stroke is that the load on the crankpin is increased (since the crankpin has less offset) with more tendency for the piston to 'rock' in the bore. The choice of bore: stroke ratio, therefore, has to be something of a compromise. Most modern engines tend to adopt near 'square' proportions, or over-square for racing purposes. Sports engines, on the other hand, are normally under-square.

It does not necessarily follow that the bore:stroke ratio employed by the engine designer is an optimum, nor are small differences between one engine and another necessarily significant. To reduce production costs two sizes of engines in a range often share the same crankcase and crankshaft, in which case they have the same stroke. Their different (displacement) sizes are obtained by using different bores.

## Scavenging

As explained in Chapter 3 there is an overlap between exhaust opening and transfer port opening on two-stroke engines, during which time the incoming (fresh) gases help push out remaining exhaust gases through the exhaust port. The usual arrangement is to have the transfer port and exhaust port diametrically opposite so that as the transfer charge fills the top of the cylinder it pushes the exhaust gases to the other side and out — Fig 12.3(A). Inevitably some of the fresh charge flows out through the exhaust as well, so this simple form of scavenging is wasteful (ie, increases fuel consumption), and the fact that a proportion of the incoming charge is lost also reduces the potential performance of the engine.

A much better arrangement is to use a piston with the top incorporating a flat section facing the intake, called a *deflector*. This is shaped to direct the incoming charge upwards in the form of a swirl or 'loop' so that most of the charge remains in the cylinder whilst still being effective in promoting scavenging action. This is known as *loop scavenging* — Fig 12.3(B). The effectiveness of such a system depends on the actual shape of the deflector. (It is also useful to remember that if an engine with a deflector piston is disassembled, the piston should be replaced the right way round — with the deflector nearest the transfer port.)

Efficient loop scavenging can only be produced in a glow engine, where the head can be suitably shaped to accommodate the deflector. This is not a practical proposition in a diesel as the 'matching' shape would have to be on the underside of the contra-piston, which would also have to be constrained against rotation. Partial loop scavenging can, however, be achieved by using

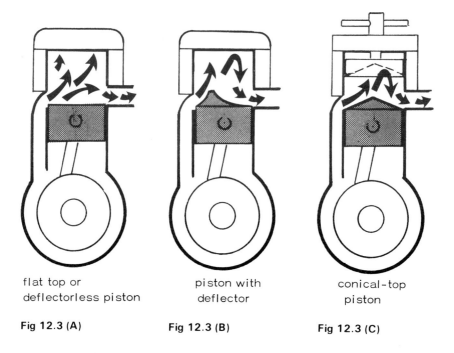

flat top or
deflectorless piston

piston with
deflector

conical-top
piston

**Fig 12.3 (A)**          **Fig 12.3 (B)**          **Fig 12.3 (C)**

a conical top piston and matching recess in the contra-piston — Fig 12.3(C). Domed pistons have also been used to produce loop scavenging, both in diesels and glow engines.

### Radial porting

All the illustrations show an asymmetric arrangement of cylinder porting — ie, exhaust port on one side of the cylinder, with the transfer diametrically opposite. This is the usual arrangement adopted for glow engines. An alternative arrangement is *radial porting* where the exhaust takes the form of a series of slots right around the circumference of the cylinder. The transfer port(s) are then cut in the cylinder walls (or cylinder liner), the tops of these ports terminating in the pillars between the individual exhaust slots. This type of porting is commonly used on diesels. There are the exceptions. Some glow engines employ radial porting, and some diesels have been produced with glow engine type porting.

An advantage of radial porting is that it enables the actual exhaust opening area to be increased without having to increase the *depth* of the exhaust port. A disadvantage of radial porting is that it is more difficult to provide effective 'loop' scavenging.

### Schnuerle porting

Various attempts have been made to improve on loop scavenging, using a flat top piston. The one which has proved most effective — and superior to conventional loop scavenging — is *Schnuerle porting*. This is now becoming an almost universal choice for high-performance glow engines.

There are many individual variations on Schnuerle porting, but basically the system employs three separate transfer ports. Two of these are positioned and shaped so that they direct the incoming charge sideways into the top of the cylinder, away from the exhaust. The third or *booster port* then opens to push this charge upwards. Provided the ports are suitably shaped, and the timing is correct, this produces very efficient scavenging without the need for a deflector on the piston. It is also superior to conventional loop scavenging as there is less mixing of inlet and exhaust gases (increasing power and reducing fuel consumption); and because of the better gas flow in the top of the cylinder, gives more uniform heat dissipation with reduced cylinder distortion due to cooling.

**Fig 12.4**

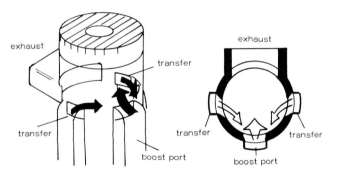

Schnuerle porting is shown in diagrammatic form in Fig 12.4. Here the two transfer ports are diametrically opposed and at 90 degrees to the exhaust (but cut at an angle to direct flow away from the exhaust. There are many variations of this (eg, the transfer ports may be much closer to the boost port), but all work on the same principle of injecting the charge into the cylinder 'sideways and upwards' on the opposite side of the cylinder to the exhaust port.

The only real disadvantage of Schnuerle porting is that it is considerably more difficult and costly to produce than conventional porting. Its application is, therefore, more or less confined to the more expensive high-performance engines.

### Squish head

Another feature which may be used on a high performance engine is a *squish head*. A conventional head shape is concave at the bottom, forming a shallow hemispherical combustion chamber (or cylinder head volume). With a squish head, the outer section of the head is flat, with a smaller hemispherical combustion chamber in the middle — Fig 12.5. The narrow flat section when the piston is at TDC forms a *squish band* which has been found to give more efficient combustion and thus improved power output. More efficient combustion improves the *mean effective pressure* in the cylinder — see Appendix 2.

**Fig 12.5**

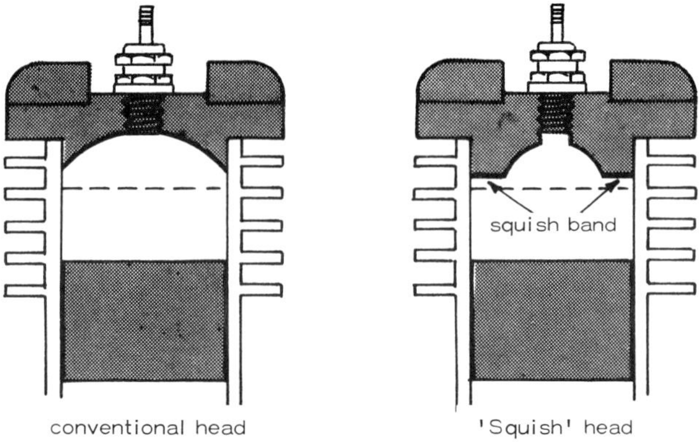

conventional head                    'Squish' head

### Piston rings

The piston on most of the larger sizes of glow engines is fitted with a ring or rings (see also Chapter 14). These may take the form of one or more conventional piston rings fitted in grooves some distance down from the top of the piston, or a *Dykes ring* fitted right at the top of the piston.

A Dykes ring is one designed to expand under pressure applied to one side and thus work as a *pressure-energised seal.* It is usually L-shaped in section and fitted in a special groove to give maximum sealing effect when the piston is being forced downwards under the pressure of the expanding gases. When this gas pressure is relieved approaching the bottom of the stroke (ie, by the exhaust opening), and for part of the upward stroke until the fresh charge starts to be compressed in the head, the ring is a relatively 'loose' fit in the bore. The resulting reduction in rubbing friction can materially improve performance on high speed engines and is generally better than the other 'standard' method of using a slightly tapered bore with a conventional piston to give freer movement over the lower part of the stroke.

# CHAPTER THIRTEEN

# CARBURETTORS

THE BASIC TYPE of carburettor employed on model engines is very simple. It consists of an *intake tube* through which is fitted a smaller tube called the *spraybar*. The spraybar has a hole at its centre and is fitted with a tapered wire or *needle valve* which can be screwed in or out from one side. The other side of the spraybar (tube) is connected to the fuel tank. The amount the tapered end of the wire (needle valve) projects beyond the central hole in the spraybar controls the effective opening through which fuel can escape into the intake tube. It thus works as an adjustable jet — see Fig. 13.1.

**Fig 13.1**

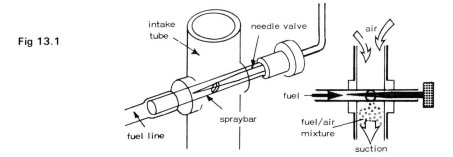

The principle on which this type of carburettor works is that when the engine rotates, air is sucked into the open end of the intake tube by the pumping action developed in the engine crankcase. This air flowing past the spraybar sucks out fuel from the spraybar hole in the form of a fine spray which becomes mixed with the air, this mixture of air and fuel then being sucked into the crankcase. The setting of the needle valve determines the amount of fuel sucked out of the spraybar, and hence the *fuel-air mixture* proportions.

The effectiveness of the suction — and 'mixing' action developed is improved if intake has a *bellmouth* entry shape. It is even more effective if the spraybar is located at the narrowest part of a *venturi* shape — see Fig 13.2. But restricting the intake tube area at this point (known as the *choke area*) has one basic limitation. It reduces the amount of fuel/air mixture that can be sucked into the engine on each stroke and thus the speed and power it can develop. High-revving, high power engines, need large choke areas; lower revving 'sports' or general purpose engines can have smaller choke areas (when improved suction characteristics usually show up in easier starting).

Fig 13.2

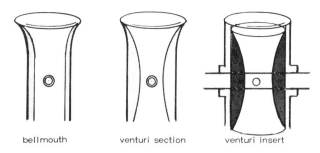

bellmouth          venturi section          venturi insert

Some engines have an intake tube with a large choke area but are also supplied with a sleeve which can be fitted in the throat section to provide a smaller choke area. In this case the full choke area is used with a pressurised fuel system, but for a conventional hook-up (ie, without pressurisation) the sleeve would be fitted to the carburettor.

Choke inserts, where supplied, are usually of true *venturi* shape to give maximum suction effect.

Suction characteristics are not helped by *forcing* air into the intake tube, eg, by facing the intake forwards. Rather the reverse. For that reason the intake tube is either fitted vertically in front of the cylinder (in practice, inclined slightly forward); or horizontally to the back of the crankcase, facing backwards — see Fig 13.3. The (near) vertical arrangement gives the equivalent of a 'downdraught' carburettor with an upright (mounted) engine and an 'updraught' carburettor with an inverted (mounted) engine.

Carburettors of this type are usually non-critical, provided the spraybar assembly is tight with no air leaks. For most efficient suction, the hole in the spraybar should face 'sideways' — see Fig 13.4.

**Fig 13.3**

**Fig 13.4**

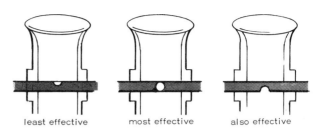

least effective    most effective    also effective

A variation on this simple carburettor design is to dispense with a spraybar as such and terminate the fuel line entry in a jet hole on one side of the intake tube at the throat position. The needle valve is mounted in a separate fitting diametrically opposite — Fig 13.5(A). Alternatively the spraybar may only extend to the centre of the choke area with fuel feed from the same side as the needle valve, as in Fig 13.5(B). This is the arrangement widely adopted in R/C carburettors which are fitted with an additional *throttle* control.

**Fig 13.5**

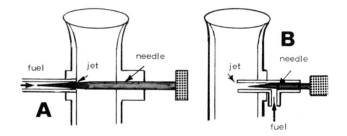

### R/C carburettors

The needle valve of a basic carburettor only adjusts the *mixture* to produce satisfactory running of the engine. The speed at which the engine runs 'two-stroking' is then governed purely by the 'load' it is driving — ie, the size of propeller. Alterations in needle valve setting will then only affect speed in two ways. If the mixture is made excessively lean (needle valve screwed in too far), the engine will become 'starved' and stop. If the mixture is made excessively

**Fig 13.6**

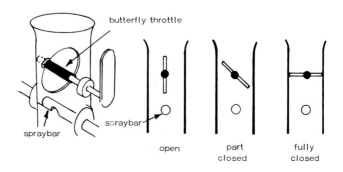

rich (needle valve screwed out from normal running position) the engine will change from 'two-stroking' to 'four-stroking' and run at some slower speed. In other words the needle valve cannot work as a *throttle* control.

Where speed control is required, as on R/C engines, a modified form of carburettor with a separate throttle device is required. The simplest form of throttle is a butterfly valve fitted in the intake tube -- Fig 13.6. Rotation of this flap alters the effective opening of the intake tube, and thus the amount of air sucked in.

This form of throttle is simple to make but has very limited response. It is also necessary to limit the movement so that the butterfly flap is never completely closed, so that air supply is never completely shut off (the alternative is to drill an air bleed hole in the flap). Without any air supply, crankcase suction would simply suck out solid fuel from the spraybar (or fuel jet), resulting in flooding of the engine.

### The barrel throttle

The standard form of throttle adopted for R/C engines is the *barrel throttle* which works on a similar principle to the butterfly valve, but without its limitations. It consists, basically, of a cylindrical 'plug' surrounding the spraybar or fuel jet but free to rotate. This cylinder is bored through from top to bottom, the fuel jet coming in the middle of this opening.

**Fig 13.7**

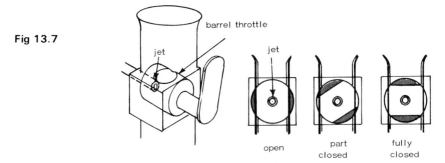

In the 'open' position (full throttle), full choke area is exposed. Rotating the barrel then progressively reduces the effective opening both at the top (for air entry) *and* bottom (for mixture delivery) — see Fig 13.7. In other words, movement of the barrel controls both air *and* mixture flow. Thus it acts as a 'true' throttle capable of varying the *quantity* of mixture delivered, providing a range of speeds over which the engine will run.

Again, to prevent flooding, it is necessary to stop the barrel movement before the opening is completely closed, so actual movement is limited by an *idle adjustment screw*. Even so, the mixture at this throttle setting will still tend to be excessively rich, so provision is also usually incorporated to provide additional air at this setting. This can be done by making the top opening of the barrel larger than the bottom one, notching the barrel to provide a separate air path, or having a separate air bleed hole drilled through the side of the intake tube. The latter is usually the best solution for an airbleed hole can be made adjustable by fitting it with a screw which enables the effective area of the airbleed hole to be altered. In effect this second adjustment is an *idling mixture control.* (Fig 13.8).

**Fig 13.8**

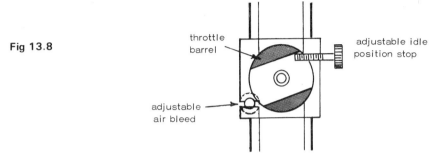

Pressurised fuel systems

**Pressurised fuel systems**
The barrel throttle still has one basic limitation for R/C aircraft. It operates with a fixed mixture jet initially adjusted under 'static' conditions (ie, model stationary). In flight the relative positions of carburettor and fuel tank will be different when climbing and diving, and the fuel flow can also be affected by 'g' forces in manoeuvres. Put simply, this can make the fuel either easier or harder to suck through the carburettor jet, which can cause mixture changes in the carburettor.

The answer to this particular problem is to *pressurise* the fuel supply. Any 'in flight' changes will then have a negligible effect. There are two obvious sources of pressure — a tapping point on the engine crankcase or one on the engine exhaust or silencer. In both cases the tapping point consists of a hole

**Fuel pressurisation pump replacing standard backplate on a front rotary engine with the pump rotor driven by the engine crankshaft.**

YS pressurisation system using crankcase pressure fed through a pressure regulator.

fitted with a nipple to which a length of flexible tubing is fitted. The other end of the tube is then taken to the vent pipe of the fuel tank. The tank filler tube must also be capped or sealed to complete a closed pressurised system (see Chapter 9).

Pressure tapped from the *silencer* is more effective than that tapped from the crankcase as exhaust pressure *decreases* with decreasing engine speed and thus provides automatic compensation for the over-richening effect of a throttle at low speeds. Crankcase pressure remains high at low speeds and tends to aggravate the over-richening effect of a throttle at low speeds. Some crankcase pressurised systems, however, include a separate pressure regulator or similar device to provide a substantially constant pressure with varying speed.

Butterfly-type throttles work better when linked to a flap closing exhaust opening as throttle is closed. This system was also used with early barrel-type throttles.

One other advantage of a pressurised system is that the amount of suction developed at the throat of the carburettor is now a minor consideration. In other words, the choke area can be increased by as much as 50 per cent without suffering any ill effects.

Some pressurised systems developed for model engines employ a small crankshaft-driven pump to produce pressure. The advantage of such a system is that the pressure developed is directly related to the speed of the engine (and also the actual pressure developed is determined by the design of the pump). Pumps of this type are fitted by replacing the standard backplate on a front rotary engine with a special backplate incorporating the pump. Examples are the Perry pump (American), Graupner HB (German) and YS pump (Japanese). They are normally intended for use with a matching design of large choke area carburettor. Pressure actually delivered to the carburettor may or may not be controlled by an adjustable pressure regulator.

### More advanced throttles

The alternative to a pressurised system is a type of throttle which itself is capable of providing automatic mixture control. The requirement, really, is to reduce the *quantity* of fuel supplied by the jet as the throttle is progressively closed.

There are a number of ways in which this can be done. One is to use a spraybar with a slit instead of a hole. A close fitting sleeve on the spraybar then moves over the slit as the throttle is closed to reduce the effective opening and thus the amount of fuel delivered from the jet — Fig 13.9. The necessary sleeve movement is produced either by moving the barrel sideways or the sleeve sideways, this motion being controlled by throttle arm movement. Alternatively, the sleeve itself can also be slotted and rotate with the barrel movement, reducing the effective jet opening as the throttle is closed. Some provision must be incorporated to adjust the sleeve position, which is then the *idle mixture control*. A separate adjustable airbleed may or may not be incorporated in throttles of this type. If it is, it forms a second idling mixture control.

**Fig 13.9**

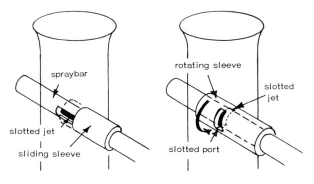

Another method is to feed fuel from the main jet through tapered grooves in the surface of the barrel and thence through a hole into the choke area. The position of the barrel determines the amount of groove opening present, which decreases as the throttle is closed.

Other designs use separate adjustable needle valves, one for the main jet and one for the idling jet. The latter jet opening is automatically decreased as the throttle is closed, thus reducing the quantity of mixture.

A point to bear in mind is that a variable jet (or idle mixture needle valve) will produce a weaker mixture if screwed in and vice versa; but an adjustable air bleed will work in the opposite sense.

### Adjusting R/C carburettors

As a general rule, with an R/C carburettor the main needle valve should be adjusted first to give smooth two-stroking with a bias to a slightly rich setting with the throttle in the fully open position. The idle mixture is then adjusted for consistent running with the throttle fully closed. (There are, however, some types of R/C carburettor which require the *slow* speed adjustment to be made *before* the high speed setting.)

In the photo on left, barrel throttle is fully open. Rotation of the barrel begins to close off the opening for air intake (centre). In the closed position (right), air supply is cut off, except for the small amount that can pass through the groove in the barrel seen on the right. Other types of barrel throttles have a separate air bleed hole in the body of the carburettor.

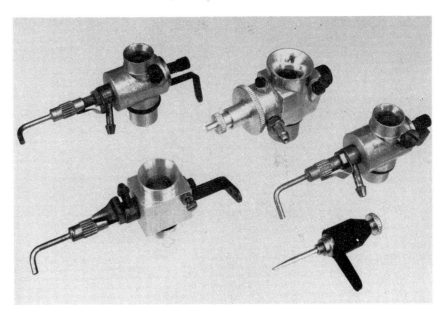

Examples of different R/C throttles. Example at the bottom right is an in-flight variable needle valve.

The position of the *throttle arm* for low speed running can be adjusted by a screw on the carburettor body. Thus two adjustments are involved in setting up the idle properly — the throttle 'stop' screw and the idle mixture control. The aim is to achieve *consistent* low speed running with no risk of the engine stalling, with immediate pick-up when the throttle is opened. To ensure consistent running it is best not to try to get the idle speed too low. On the other hand the idling rpm must be low enough for the model to lose height in a 'powered glide'.

In practice, it is best to set the throttle 'stop' position by the controlling *servo* movement, backing the throttle stop screw right off. Adjustment of movement should be made with the *throttle trim control* on the transmitter fully forward. Closing the throttle lever will then give idling speed. With the throttle in this position, moving the throttle trim lever back will then stop the engine.

The more experienced R/C modeller may prefer a more sophisticated adjustment, adjusting the idle 'stop' position with the throttle trim control central. Forward movement of the throttle trim lever on the transmitter will then give a small increase in rpm (useful on a landing approach); and moving the throttle trim lever back will stop the engine.

**Needle valve control**

On certain types of radio control models —eg, pylon racers — throttle control is less essential than an ability to re-adjust the actual running mixture in flight. This can be particularly useful with 'racing' engines employing carburettors with large choke areas where mixture adjustment can be quite sensitive.

In this case the carburettor is of the simple non-throttle type but with a modified needle valve unit which can be adjusted in flight via a lever operated from a servo.

**Fuel filters**

R/C type carburettors, and automatic carburettors in particular, depend on fine 'metering' openings for idle mixture control. These can easily become clogged by any solid impurities in the fuel. It is thus highly advisable to use a fuel filter in the line between tank and carburettor, and also to use a separate filter when filling the tank from a fuel container. If the carburettor has slit-type jet openings, then the use of at least a fuel line filter is essential.

With plain carburettors (ie, non-throttled types), a filter can be used in the fuel line but is normally not necessary.

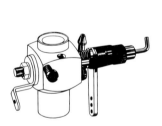

**Kavan in-flight adjustable needle valve.**

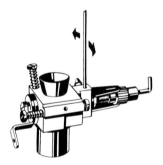

**Kavan automatic carburettor.**

**Proprietary 'automatic' R/C carburettors**

Individual manufacturers have developed — and continue to develop — R/C carburettors providing automatic adjustment of mixture over the lower end of the throttling range. In the main these are produced by engine manufacturers to match their particular engines, although other types (such as the Kavan and Perry carburettors) are used with, or fitted to, a variety of different engine makes. The following are brief notes on the majority of the current automatic R/C carburettor types.

*MERCO*

Sideways-moving barrel type with single jet and main needle valve mixture adjustment. Second adjustable needle valve enters open end of jet tube as throttle is closed to reduce quantity of fuel leaving the jet.

*HP (original)*

Two needle valves with same operating principle as above.

*TAIPAN*

Sideways-moving barrel type with single jet. Two needle valves with same operating principle as above.

*OPS*

Sideways-moving barrel type with single jet. Two needle valves with same operating principle as above.

*WEBRA (TN)*

Sideways-moving barrel type with single jet. Two needle valves with same operating principle as above.

*FOX*

Separate jets for high speed and idling. Modified barrel with groove-controlled delivery of fuel to main jet; only idling jet operative at idling speed.

*KAVAN*

Barrel-type throttle with full length spraybar with radial slit for jet, encircled by sleeve with circular hole. Sleeve rotates with closing throttle to reduce length of jet. Single needle valve control of main mixture. Adjustable sleeve setting. Separate adjustable airbleed for idling mixture setting.

*PERRY*

Barrel-type throttle with special mixture-control and transfer chamber with slip ports. Single needle control of main mixture. Knurled disc adjustment of idling mixture.

**Perry automatic carburettor.**

## OS

Barrel-type throttle with full length spraybar with longitudinal slit for jet. Sideways sliding sleeve on spraybar reduces jet opening as throttle is closed, with adjustment by screwdriver. Single needle valve controlling main mixture.

## SUPER TIGRE

Twin jet sideways-moving barrel-type with full length spraybar having a longitudinal jet slit. Idle needle is rod-shaped, fitting inside spraybar, reducing jet opening as throttle is closed. Main mixture control by second needle valve.

(1)   needle
(2)   O-ring
(3)   carb. nut
(4)   lock nut
(5)   spray bar
(6)   cam screw
(7)   barrel valve spring
(8)   barrel valve

(9)   carb. arm
(10)  low speed needle
(11)  carb. body
(12)  Fuel line nipple
(13)  screw set
(14)  needle valve assy
(15)  retainer
(16)  Venturi

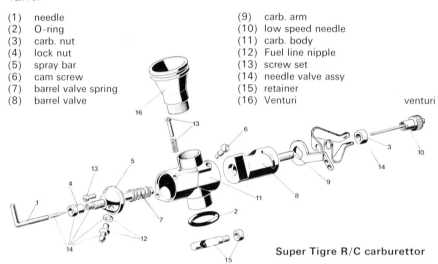

**Super Tigre R/C carburettor**

## ENYA G-type

Special barrel with grooves reducing fuel flow to single jet as throttle is closed. Single needle valve control of main mixture. Separate airbleed for adjustment of idling mixture.

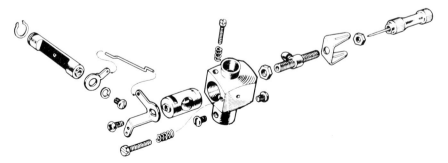

**Exploded drawing of Enya G-type throttle.**

*HP (FM)*

Barrel-type with unrestricted choke area and with main jet at side controlled by main needle valve. Separate idling mixture screw controls mixture over lower end of throttle range (and also acts as a throttle stop). Separate air bleed for adjustment of idling mixture.

*HP (61-FS)*

Sophisticated barrel-type incorporating special metering valve in transfer chamber (similar to Perry carburettor). Needle valve for main mixture adjustment. Metering valve adjustment by knurled disc.

# CHAPTER FOURTEEN

# ENGINE CONSTRUCTION

THE MAJORITY of model engines follow a more or less standardised form of construction — a crankcase casting in light alloy with steel cylinder liner and light alloy head. The crankcase casting may extend the full length of the liner (with cast-in fins for cooling), or the finned section be in the form of a separate machined jacket. There are exceptions. In the case of small glow motors the complete cylinder may be machined from mild steel, with integral finning.

The crankshaft bearing housing may also be incorporated in the crankcase casting, or be a separate casting. The back of the crankcase may be integral with the crankcase, or again a separate casting. Assembly of the individual external components is by bolts.

Enya 60 engine shown alongside second engine disassembled.

Part-sectioned engine shows clearly the cylinder liner, cylinder jacket and head as separate units; also the crankshaft carried by two ball races.

The piston is either of cast iron (if a plain type) with a lapped fit; or aluminium, fitted with rings. Diesel engine cylinders or cylinder liners are hardened. Glow engine liners may or may not be hardened, but are usually unhardened.

Ringed pistons do set one constructional problem. If the engine has large port openings in the cylinder (liner), there is a distinct possibility of a piston ring 'expanding' into a port opening and being damaged. To prevent this from happening port openings are bridged, ie, made in the form of a series of side-by-side openings of relatively short circumferential length instead of one large opening.

With larger glow motors there is a trend to adopt a chrome plated brass liner with a plain aluminium piston. The combination of aluminium and brass is one with similar expansion characteristics, thus maintaining a good piston fit over a range of running temperatures. Chrome plating of the brass liner produces a hard surface to resist wear. This form of construction is known as ABC (aluminium-brass-chromium). Engines of ABC construction may also be fitted with piston rings.

Another feature of engines with ABC construction is that due to the 'matched' expansion of piston and cylinder liner, piston and cylinder fit can be relatively 'loose' to start with so that very little running-in time is needed.

In fact, some ABC engines may be effectively run-in after just two or three runs on a rich mixture. The manufacturer's instructions in this case will probably state that the engine is suitable for 'in flight' running-in, implying that no preliminary running-in is required at all before the engine is installed in a model.

Other material combinations have been tried, of which only AAC or aluminium piston in aluminium cylinder with chrome plated bore has achieved any degree of success. ABC remains the most favoured modern choice and has shown improved performance over conventional constructions, but at the expense of some loss of ease of operating for starting and running adjustment.

Crankshafts are invariably of hardened steel, finished by grinding. Fully hardened (or through hardened) crankshafts tend to be brittle and break readily on impact (eg, in a crash landing), so it is desirable to employ heat treatment of a hardened shaft to relieve some of the hardness and thus the brittleness.

The crankshaft may be carried in a plain bearing, in which case the bearing is usually a bush of cast iron or sintered bronze, cast in or press fitted to the crankcase and honed to size. A typical crankcase alloy can provide quite a good bearing on its own, and some smaller engines may be made on this basis (ie, without a main bearing sleeve). The crankshaft fit and bearing finish of a

Cox engine construction favours crankcase machined from solid extrusion (rather than cast), and machined steel cylinder with integral finning. This engine is radially mounted to its own tank.

Unusual type of finned head on this Graupner HB engine is to provide satisfactory air cooling in helicopters where there is no slipstream blowing over the engine.

plain bearing engine has a considerable effect on the performance of an engine. A tight bearing may take a very long time to run in properly. A poor fit and/or finish can lead to loss of performance and a high rate of bearing wear.

The use of ball races instead of a plain bearing substantially reduces running friction, and thus improves the performance of any engine. A ball race engine should always be faster revving than its plain bearing counterpart, and bearing life should be longer. All high performance engines are fitted with twin ballraces for the crankshaft bearing.

*Ball* bearings are superior to *roller* bearings in such applications since they have less running friction. However, needle-type roller bearings may sometimes be used in an attempt to reduce friction between the con rod big end and crankpin. Any advantage is debatable since the friction of the small needles necessary is unlikely to be appreciably less than that generated by a plain bearing.

# CHAPTER FIFTEEN

# MARINE ENGINES

THE MAJORITY of model marine engines are watercooled versions of aero engines. The only difference is that the finned cylinder jacket is replaced by a larger, hollow jacket through which cooling water can be circulated. A flywheel is also necessary and nearly all marine engines are supplied complete with matching flywheel.

The usual arrangement for providing a supply of cooling water to the head is from a scoop fitted to the bottom of the hull immediately behind the propeller. This is connected by plastic tube to the *bottom* tube in the head. The top tube in the head is then taken, via another length of plastic tube to an outlet in the side of the hull, or the transom — Fig 15.1. This outlet can simply be a hole in the hull through which the tube is a tight fit, or a special fitting called a transom flange.

One manufacturer's range of watercooled marine engines with standard and R/C carburettors.

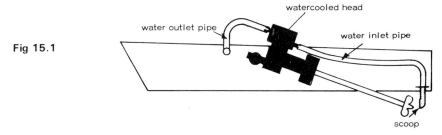

Fig 15.1

Engine installation is shown in Fig 15.2. The most important requirement is that the engine crankshaft and propeller shaft should be in line, which normally means that the engine mount is angled. The mount itself usually consists of two substantial blocks of solid hardwood cut to the required taper and securely fastened in the hull. To this is screwed a thick laminated plastic (Tufnol) plate (or a dural plate), cut out to receive the engine. The engine is then bolted to this plate via its beam mounting lugs. Note that a marine engine is always mounted facing 'backwards'.

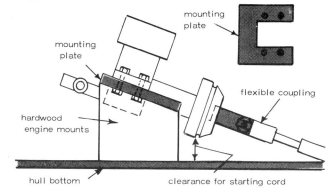

Fig 15.2

Connection between the engine crankshaft and propeller shaft is by means of a *flexible coupling*. There are many different types of flexible couplings and choice of any particular type is not critical provided it is of suitable strength (ie, specified as suitable for the size of engine), and the coupling *threads* match those of the engine crankshaft and propeller shaft used. One part of the coupling screws on to the end of the engine crankshaft, and the other to the propeller shaft. It does not matter which way round the coupling is fitted.

The fact that flexible couplings can transmit the drive through an angle does *not* mean that alignment between engine crankshaft and propeller shaft is not important. The better the alignment, the less the vibration of the drive. If it *is* necessary to transmit the drive through an angle for any reason, then a *double coupling* should be used. This is a coupling with two 'end' components and an intermediate member.

To complete the installation the engine is fitted with a silencer, which may be of special type, and/or require the use of a separate manifold in order to connect the silencer to the engine exhaust. The outlet pipe of the silencer is fitted with a length of large diameter tubing to direct the exhaust overboard, either through the side or through the transom — Fig 15.3. This piping can be metal (eg, copper tube), or more conveniently silicone tubing.

A second silencer can be included in this outlet pipe (to improve silencing,

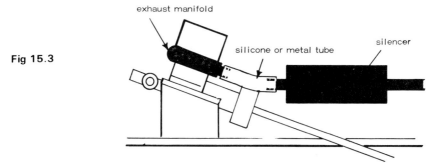

**Fig 15.3**

at the expense of some increase in power loss); also an oil trap to collect 'liquid' exhaust waste. The latter may be an 'anti-pollution' requirement on some ponds, although a simple silencer itself can also work as an oil trap. Remember that silencers and oil traps need draining periodically.

The fuel tank can be located in any convenient place in the hull, but must be at a level where the top of the tank is no higher than the spraybar on the engine carburettor (see Chapter 14 and Fig 15.4).

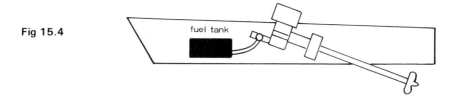

**Fig 15.4**

**Starting**

Marine engines are started by a cord (ideally a leather thong) passed around the groove in the flywheel and given a sharp pull upwards to spin the flywheel over rapidly — Fig 15.5. This requires a certain knack. It needs two hands, one holding each end of the starting cord, gripping the hull of the boat between the knees. The left hand keeps tension on the cord as the right hand pulls sharply upwards — releasing the 'left hand' end of the cord at the appropriate moment to allow it to be pulled free of the flywheel.

Starting is much easier if an electric starter is used. In this case the starting

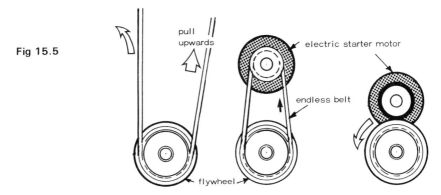

**Fig 15.5**

'cord' is in the form of a synthetic rubber belt to provide a 'pulley drive' between starter and flywheel. When running, the belt is just left to lie slack, clear of the flywheel. (Remember a belt must be assembled over the drive before coupling up.) Alternatively a starter may be fitted with a friction drive drum or wheel which can be applied directly to the top of the flywheel. In this case no starter belt is needed.

Starting adjustments are similar to those of aero engines (see Chapter 4), except that the initial adjustment for running should be left 'four-stroking' (or on a low throttle setting in the case of an R/C engine). This is because the flywheel and marine propeller will provide very little 'load' when the propeller is out of the water. Final adjustment of the needle valve is then made when the model is in the water. Thus if the mixture setting is leaned out, the engine will over-rev.

This will need to be a fairly 'rich' setting, as the engine will gain a lot of speed and 'lean out' once the boat has accelerated up to its steady speed. If a starter is used, it is possible — and often more convenient — to start the engine *with* the model in the water.

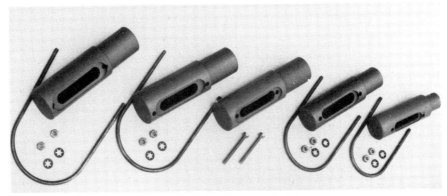

**Examples of marine engine manifolds by Ripmax Models.**

Inlet pipe is the lower one on the jacket, with outlet pipe usually on the opposite side.

## Marine propellers

Unlike aero engines there are no specific matching prop sizes recommended for individual engines. Also there is a much greater variety of different propeller shapes. As a general rule, 2-bladed propellers are best for 'racing' performance; and 3-bladed propellers for other applications. The following is then a *rough* guide to propeller selection. For 'racing' performance it will be necessary to experiment with other sizes, and different makes, in order to get the best out of the particular engine-hull combination.

| Engine size | | | | |
|---|---|---|---|---|
| cc | cu/in | | | |
| .5 | — | 1¼ × 1½ | 1¼ × 1¾ | 1¼ × 1¼ |
| .75 | .049 | 1¾ × 1¼ | 1½ × 1½ | 1¼ × 1½ |
| 1 | — | 1¾ × 1¼ | 1½ × 1½ | 1¼ × 1½ |
| 1.5 | .09 | 1½ × 2 | 1½ × 2 | 1¼ × 1½ |
| 2.5 | .15 | 1½ × 3 | 1½ × 1¾ | 1¼ × 1½ |
| 3.5 | — | 1½ × 3 | 1½ × 1¾ | 1¼ × 1½ |
| | .19 | 1½ × 3 | 1½ × 1¾ | 1½ × 1½ |
| | .29 | 2 × 3 | 1¾ × 2 | 1½ × 1¾ |
| | .35 | 2½ × 3 | 2½ × 2½ | 2¼ × 2¼ |
| | .60 | 3 × 2¾ | 2½ × 3 | 2½ × 2½ |

In these sizes the first figure is the diameter (in inches) and the second figure is the pitch (in inches).

Marine engines are usually watercooled versions of their aero engine counterpart. Watercooling jacket may include the cylinder head, or a separate finned head may be used. The conical or 'low profile' form of flywheel is now more or less universal.

Typical marine glow engine with manifold fitted to the exhaust stub.

Heavy duty flexible coupling (Ripmax 'Flexidisc') intended for use with extra powerful or large marine engines.

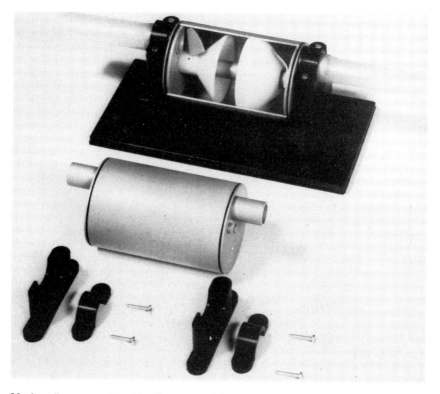

Marine silencer produced by Graupner with its mounts and (above) special trap to be connected in series with the silencer to collect oil from the exhaust, thus minimising pollution.

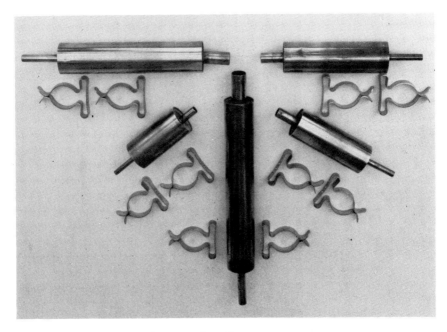

Rip-max marine silencers with mounting clips.

Examples of different types of couplings from the Ripmax range.

# CHAPTER SIXTEEN

# CARE AND MAINTENANCE

A MODEL ENGINE is a precision-made machine and should be treated as such. No parts should ever be gripped in a vice, even for disassembly. At best it will mark the surface of the part. At worst, it will distort and ruin it. Nor should a screwdriver or similar object ever be pushed through the exhaust port opening to 'lock' a piston in place to lift a liner or cylinder jacket during disassembly. In fact, taking an engine apart at all should be avoided, unless absolutely essential to replace a broken or damaged internal part.

These are extreme cases of abuse, but not all that rare. Almost as bad is turning an engine over after a bad crash landing to check that it is still 'free'. At least clean off any dirt, etc, from the outside of the engine, and particularly the exhaust port (if this is not covered by a silencer), and carburettor intake. In fact, always keep the outside of an engine clean. Wipe it over with an oily rag after each session of use. Squirt a little light oil (or some fuel) into the carburettor (and exhaust as well, if easily accessible), and turn the engine over a

Always use a matching spanner for tightening or removing a propeller nut — never pliers.

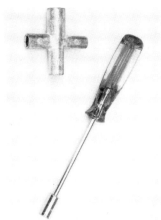

Universal spanner fits a variety of prop nut sizes and matches glow plugs (6 mm hexagon). Shown also is a special glowplug spanner. Right: It is even more important to use a proper spanner on glow plugs. A universal spanner may not clear the head fins on all engines, when a slimmer spanner must be used.

number of times to disperse this lubricant through the engine. This will provide protection against corrosion (eg, from acidic combustion products which may otherwise be left in the engine) whilst it is not being used. If laying up an engine for a long period, be even more generous with the oil and crank over several dozen times to ensure complete distribution. Remove the silencer and tape up the exhaust port opening and the carburettor intake before storing away in a clean, dry place.

Apart from dirt getting inside the engine, the main cause of engine wear and loss of power is vibration. Make sure that the nuts on the engine mounting screws are tight and re-check periodically. If using plain nuts use *two* nuts on each screw. Self-locking nuts are better. Use a propeller which is properly balanced to minimise vibration (see Chapter 6).

The major components of the engine are usually held together with screws, and screws do work loose. Check the cylinder head screws and crankcase fixing screws periodically. Make sure that they are tight, but do not over-tighten as this could strip the thread in the casting. Loose screws can mean air leaks and loss of performance.

Always use the correct tools for any particular job — the right size of Phillips screwdriver for socket-head screws, and a plug spanner for removing or replacing a glow plug (never pliers).

If replacement parts are ever needed (apart from glow plugs), always obtain the correct spare(s) for that particular engine. Some parts, too, will only be available in matching sets (eg, pistons and cylinder liners). The replacement of ball races may be tricky, and even impossible for the average owner if shrink-fitted originally. Thus major replacements are usually better left for the service agent for the particular engine to do.

A point to remember with glow engines using methanol fuels is that methanol readily absorbs moisture from the air. If an engine is stopped after a 'wet' run—e.g. running rich, such as when throttled back in landing an R/C aircraft—unburnt fuel remaining in the engine can cause rusting of internal metal parts if the model is put away in this state. It is thus safest to run an engine dry at the end of a flying session (or a bench run) by running at full throttle and shutting off the fuel supply (e.g. by disconnecting the fuel line) to stop the engine.

# ENGINE SIZES

CONVENTION HAS IT that glow engine sizes are always given in *cubic inches* (cu in) displacement and diesel engines in cubic centimetres (*cc*) displacement, regardless of country of origin. The bulk of the model engines produced also fall into 'standard' sizes. There are:

Glow engines (cu in): .049, .09, .15, .19, .29, .35, .45 and .60.

There are different sizes produced, notably smaller sizes than .049 cu in, and also a tendency by some manufacturers to 'fill in' between these standard sizes with intermediate sizes — eg, .12, .25 and .40.

Larger glow engines are also now produced in limited numbers to match the power requirements of larger aircraft. The only 'standard' size to emerge so far is .90 cu. in. (15cc).

Production of diesels larger than 3.5cc displacement is largely non-existent, although diesel conversions of glow engines have proved practical from .049 size up to .60 cu. in.

Spark-ignition engines, when originally in vogue, had displacements quoted in cc in Britain and Europe and those manufactured in the USA in cu. in. The present-day reintroduction of spark-ignition engines for giant size models is largely based on American chain saw engine prototypes, with sizes invariably quoted in cu. in. There is no 'standard' range of sizes for these engines, although a popular choice is 2.1 cu. in. (34cc).

Engine sizes are commonly referred to by number only — eg, 049, 09, 15, etc — it being taken as understood that the number will also identify whether the engine in question is a glow engine or a diesel. The .049 engine size is also well known as ½-A.

To convert from cubic inches (cu in) to cc, and vice versa, to compare glow engine and diesel sizes, the following approximate figures should be easy to remember:

.60 cu in is equal to 10 cc
.15 cu in is equal to 2.5 cc
1 cc is equal to .06 cu in
0.5 cc is equal to .03 cu in

For more accurate conversion, use the Tables given in Appendix 3.

# PERFORMANCE CURVES AND HOW TO UNDERSTAND THEM

ENGINE TEST REPORTS appear regularly in the monthly model journals, together with *performance curves* for the particular engine under test. These curves show how *BHP* (brake horse power) and *torque* vary with engine speed or rpm. It is really only the BHP curve which is of interest, although it is the torque which is actually *measured* during a test.

Torque is the actual 'turning effort' developed by the crankshaft, due to the pressure of the fuel mixture 'fired' in the cylinder at the top of each compression stroke. It can be measured by a dynamometer in force-distance units, eg, ounce-inches usually. A torque of, say, 40 ounce-inches is equivalent to the 'turning force' generated by a 1 ounce weight on the end of a 40 inch lever arm (or a 40 ounce weight on the end of a 1 inch lever arm). This is about the sort of torque value developed by a .29 - .35 glow engine.

The *torque* developed tends to *decrease* with increasing rpm. Generally torque will be at a maximum at some moderate rpm, then gradually decrease — hence a torque curve normally tends to slope downwards from left to right. This is because as rpm increases, friction also increases absorbing any more of the available 'turning effort' developed by the engine.

*Power* cannot be measured directly, but is determined as the *product* of torque and rpm. A formula which is accurate enough for all purposes is:

$$\text{Power, or BHP} = \frac{\text{torque in ounce-inches x rpm}}{1,000,000}$$

The BHP curve tends to go on rising with increasing rpm, because the decrease in torque with speed is fairly gradual. If the engine develops say just over 30 ounce-inches torque at 6,000 rpm it may well be still developing 28 ounce-inches of torque at 12,000 rpm. At 6,000 rpm the engine is actually developing 0.2 BHP; and at 12,000 rpm it is developing 0.35 BHP.

There will come a point, however, at which the torque is falling off more rapidly so that the product of torque and rpm (ie, BHP) actually starts to fall. Thus the BHP curve continues to rise to a maximum, then falls off — Fig A.1. This is the maximum BHP for that particular engine (on the particular fuel used and under any other conditions which can affect performance, such as the fitting of a silencer). And that maximum BHP will occur at a specific rpm figure, called the 'peak' revs. If engine rpm is increased (eg, by fitting a smaller prop), both torque and BHP will continue to fall. Carried to the extreme, both torque and power will fall to zero at some much higher speed (equivalent to the speed the engine could reach running after shedding the propeller). At this speed, all the 'turning effort' produced by the engine is being used up, in overcoming internal friction.

To get an engine to develop maximum power, it must be 'loaded' to run at the speed corresponding to maximum BHP. This is done by fitting a suitable size of propeller (see Chapter 6). Test runs with a series of different propeller

**Fig A.1**

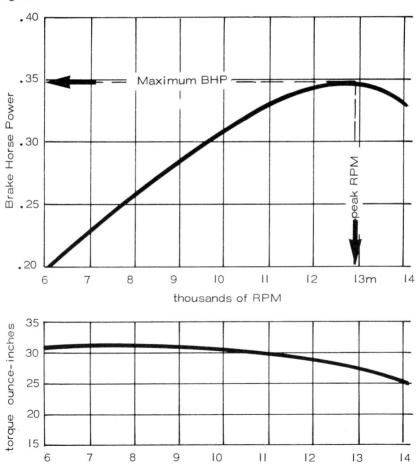

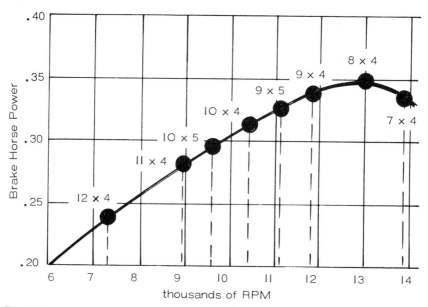

**Fig A.2**

sizes for an engine with a similar BHP curve to that shown in Fig A.1, and measuring the rpm with a tachometer in each case, might result in rpm figures like those shown in Fig A.2 (superimposed on the BHP curve). In this particular example, an 8 x 4 prop is the one which allows the engine to attain 'peak' revs. Anything larger than an 8 x 4 prop will result in slower running. A smaller prop (eg, the 7 x 4) gives *more* rpm but *less* power, because the engine is now operating past its peak.

This would be a very useful way of finding a matching propeller size, but for one other vital factor. Propeller speed measurements normally have to be made under static conditions which exaggerate the 'loading' effect of the propeller. Under flight conditions, the actual loading imposed by the propeller will decrease and the engine will speed up (see Chapter 6 again for a fuller explanation of this effect). Basically, therefore, a 9 x 4 or even a 9 x 5 prop might be the best choice in this particular example, depending on the type of model.

This example does, however, give reality to the significance of the BHP curve. (The torque curve can be ignored for most practical purposes.) It also implies that a good indication of the comparative performance of two different engines is the rpm figure it will achieve on the same propeller, but this is not an entirely reliable guide with static tests, particularly if a high pitch propeller is involved.

The obvious way to increase the maximum BHP of a particular engine is to maintain the torque as far as possible as rpm is increased. This means getting more 'turning effort' out of the engine, which can be done by using a 'hotter' fuel and/or increasing the amount of fuel introduced with each charge. The former is something which can be tried with any engine. The latter is a feature of engine design, in increasing *port areas*, starting right at the carburettor with a larger choke area; and increasing the efficiency of the scavenging (see Chapter 13). That is why some engines develop a much higher maximum BHP than other engines of similar size. They are designed to achieve a high 'peak' rpm figure.

Engine test data may also refer to *brake mean effective pressure* (BMEP) and *mean piston speed*. These are parameters related to the traditional 'basic' formula for engine performance, which is

$$\text{horsepower (per cylinder)} = \frac{PLAN}{396,000}$$

where  P  =  mean effective pressure developed by
ignition of the fuel charge, in $lb/in^2$
L  =  length of stroke, in inches
A  = piston area, in square inches
N  =  revs per minute (rpm)

Since it is virtually impossible to measure P directly, it is calculated back from the horsepower figure derived by dynamometer test. This calculated value of P is known as the *brake* mean effective pressure (as being derived from Brake Horsepower). In fact, P depends on the weight of air taken into the cylinder, the weight and quantity of fuel burned in it, and the efficiency of combustion — all of which are interrelated factors.

Mean piston speed, or more correctly *maximum mean piston speed* is calculated simply as equal to L x N divided by 6, and designated S, ie:

$$S = \frac{LN}{6} \text{ or } \frac{L \times rpm}{6}$$

The main significance of this parameter is that L and N are both subject to practical limits, and the product LN is limited by piston, ring(s) and liner properties.

# CONVERSION TABLES

## CUBIC INCHES TO CUBIC CENTIMETRES

| cu. in. | 0·0 | 0·01 | 0·02 | 0·03 | 0·04 | 0·05 | 0·06 | 0·07 | 0·08 | 0·09 |
|---|---|---|---|---|---|---|---|---|---|---|
| 0·0 | — | 0·16 | 0·33 | 0·49 | 0·66 | 0·82 | 0·98 | 1·15 | 1·31 | 1·47 |
| 0·1 | 1·64 | 1·80 | 1·97 | 2·13 | 2·29 | 2·46 | 2·62 | 2·79 | 2·95 | 3·11 |
| 0·2 | 3·28 | 3·44 | 3·61 | 3·77 | 3·93 | 4·10 | 4·26 | 4·42 | 4·59 | 4·75 |
| 0·3 | 4·92 | 5·08 | 5·24 | 5·41 | 5·57 | 5·74 | 5·90 | 6·06 | 6·23 | 6·39 |
| 0·4 | 6·55 | 6·72 | 6·88 | 7·05 | 7·21 | 7·37 | 7·54 | 7·70 | 7·87 | 8·03 |
| 0·5 | 8·19 | 8·36 | 8·52 | 8·69 | 8·85 | 9·01 | 9·18 | 9·34 | 9·50 | 9·67 |
| 0·6 | 9·83 | 10·00 | 10·16 | 10·32 | 10·49 | 10·65 | 10·82 | 10·98 | 11·14 | 11·31 |
| 0·7 | 11·47 | 11·63 | 11·80 | 11·96 | 12·13 | 12·29 | 12·45 | 12·62 | 12·78 | 12·95 |
| 0·8 | 13·11 | 13·27 | 13·44 | 13·60 | 13·77 | 13·93 | 14·09 | 14·26 | 14·42 | 14·58 |
| 0·9 | 14·75 | 14·91 | 15·08 | 15·24 | 15·40 | 15·57 | 15·73 | 15·90 | 16·06 | 16·22 |
| 1·0 | 16·39 | — | — | — | — | — | — | — | — | — |

## CUBIC CENTIMETRES TO CUBIC INCHES

| c.c. | 0·0 | 0·1 | 0·2 | 0·3 | 0·4 | 0·5 | 0·6 | 0·7 | 0·8 | 0·9 |
|---|---|---|---|---|---|---|---|---|---|---|
| 0 | — | 0·006 | 0·012 | 0·018 | 0·024 | 0·031 | 0·037 | 0·043 | 0·049 | 0·055 |
| 1 | 0·061 | 0·067 | 0·073 | 0·079 | 0·085 | 0·092 | 0·098 | 0·104 | 0·110 | 0·116 |
| 2 | 0·122 | 0·128 | 0·134 | 0·140 | 0·146 | 0·153 | 0·159 | 0·165 | 0·171 | 0·177 |
| 3 | 0·183 | 0·189 | 0·195 | 0·201 | 0·207 | 0·214 | 0·220 | 0·226 | 0·232 | 0·238 |
| 4 | 0·244 | 0·250 | 0·256 | 0·262 | 0·269 | 0·275 | 0·281 | 0·287 | 0·293 | 0·299 |
| 5 | 0·305 | 0·311 | 0·317 | 0·323 | 0·330 | 0·336 | 0·342 | 0·348 | 0·354 | 0·360 |
| 6 | 0·366 | 0·372 | 0·378 | 0·384 | 0·391 | 0·397 | 0·403 | 0·409 | 0·415 | 0·421 |
| 7 | 0·427 | 0·433 | 0·439 | 0·445 | 0·452 | 0·458 | 0·464 | 0·470 | 0·476 | 0·482 |
| 8 | 0·488 | 0·494 | 0·500 | 0·507 | 0·513 | 0·519 | 0·525 | 0·531 | 0·537 | 0·543 |
| 9 | 0·549 | 0·555 | 0·561 | 0·568 | 0·574 | 0·580 | 0·586 | 0·592 | 0·598 | 0·604 |
| 10 | 0·610 | — | — | — | — | — | — | — | — | — |

# ENGINE SPARES AND SERVICING
# —WHAT TO DO

FEW MODEL shops carry engine spares in stock. (Stocking problems would be enormous. Each engine may have anything from 20 to 30 different parts available as spares. Multiply this by the number of different engines in a typical range and the spares list runs to hundreds of different items). The local model shop is normally the place to go *for* spares, however, as he can then order and obtain them from the manufacturer or distributor concerned.

This assumes that only simple, easily fitted replacement items are required —such as a lost prop nut, or a bent needle valve. For anything requiring more than simple physical replacement which can be tackled by the non-expert, the engine should be returned to a specialist for servicing and repair as necessary.

Some engine manufacturers and importers/distributors of foreign made engines have their own service department for handling such jobs. Others do not. In that case it is necessary to send the engine to one of the firms or individuals who specialise in engine servicing. Some of these handle only certain makes of engines; others provide a service for almost any engine. A source of names and addresses of such firms is a current issue of one of the modelling magazines, such as *Aeromodeller*; but the appearance of an *advertisement* offering such a service is no guarantee of satisfactory work (although it usually is).

The following are particularly recommended:

JOHN HETER, The Haven, Rixey Park, Chudleigh, Devon. (Specialising in the servicing and repair of virtually all makes and sizes of engines).
COBRA MODEL ENGINEERING LTD, 497 Hertford Road, Enfield, Middlesex EN3 5XH. (Specialising particularly in four-stroke model engines).

# MODEL ENGINE MANUFACTURERS, IMPORTERS AND DISTRIBUTORS

## MAJOR BRITISH ENGINE MANUFACTURERS

DAVIES-CHARLTON LTD, Hills Meadow, Douglas, Isle of Man.
IRVINE ENGINES, Unit 2, Brunswick Industrial Park, Brunswick Way, New Southgate, London N11 1JL.
PAW (diesels), Progress Aero Works, Chester Road, Macclesfield, Cheshire.
MERCO, D. J. Allen Engineering Ltd, 30 Lea Valley Trading Estate, Edmonton, London N18.
TELCO SYSTEMS, Station Road, East Preston, Littlehampton, West Sussex BN16 3AG.

## MAJOR IMPORTERS/DISTRIBUTORS IN THE UK

(Country of origin of engine shown in brackets).
AUSTRO-WEBRA (Germany): Ripmax Ltd, Ripmax Corner, Green Street, Enfield, Middlesex EN3 7SJ.
COX (USA): A. A. Hales Ltd, PO Box 33, Hinckley, Leicestershire.
Ripmax Ltd, Ripmax Corner, Green Street, Enfield, Middlesex, EN3 7SJ.
ENYA (Japan): Ripmax Ltd, Ripmax Corner, Green Street, Enfield, Middlesex EN3 5BJ.
FOX (USA): Fox Manufacturing Co (UK), The Haven, Rixey Park, Chudleigh, Devon.
Model Aircraft (Bournemouth) Ltd, Norwood Place, Bournemouth.
HB (Germany): Irvine Engines, Unit 2, Brunswick Industrial Estate, Brunswick Way, New Southgate, London N11 1JL.
HP (Australia): Ripmax Ltd, Ripmax Corner, Green Street, Enfield, Middlesex EN3 7SJ.
K&B (USA): Irvine Engines, Unit 2, Brunswick Industrial Estate, Brunswick Way, New Southgate, London N11 1JL.
OPS (Italy): Irvine Engines, Unit 2, Brunswick Industrial Estate, Brunswick Way, New Southgate, London N11 1JL.
FUJI (Far East): MacGregor Industries Ltd, Canal Estate, Langley, Slough, Berks SL3 6EQ.
TIGRE (SUPER TIGRE) (Italy): World Engines Ltd, 97 Tudor Avenus, Watford, Herts.
Micro-Mold, Station Road, East Preston, Littlehampton, West Sussex BN16 3AG.
OS (Japan): OS Products Ltd, Unit 2, Brunswick Industrial Estate, Brunswick Way, New Southgate, London N11 1JL.
TAIPAN (Far East): Model Aircraft (Bournemouth) Ltd, Norwood Place, Bournemouth.